WIRING
A HOUSE

WIRING
A HOUSE

REX CAULDWELL

The Taunton Press

COVER PHOTO: **Roe A. Osborn**

The Taunton Press
Inspiration for hands-on living™

10 9 8 7 6
Printed in the United States of America

Taunton's For Pros / By Pros®: Wiring a House was originally
published in 1996 by The Taunton Press, Inc.

For Pros / By Pros® is a trademark of The Taunton Press,
Inc., registered in the U.S. Patent and Trademark Office.

The Taunton Press, Inc., 63 S. Main Street,
PO Box 5506, Newtown, CT 06470-5506
e-mail: tp@taunton.com

Library of Congress Cataloging-in-Publication Data

Cauldwell, Rex.
    For Pros / By Pros®: Wiring a House / Rex Cauldwell.
        p.   cm.
        Includes index.
    ISBN 1-56158-332-4 (paperback)
    1. Electric wiring, Interior. I. Title.
TK3285.C38   1998   96-22821
621.319'24 – dc20   CIP

To my wife, Diana, who kept the household together while I wrote.
And without the skilled help from my granddaughter, Katy,
who was always demanding to draw happy faces on the computer,
and without the help of my three cats, Little Crazy Horse, Peaches and Zipper,
who were always lying on the keyboard and monitor, I could have
finished this book at least four times faster.

ACKNOWLEDGMENTS

No one person creates a book. It is the accumulation of many hands and minds. At The Taunton Press, I would like to thank Julie Trelstad, who led me through the creation process, and Tom McKenna, who led me through the editing process. I cannot imagine two better people to work with—they made the book enjoyable. For the artwork, I would like to thank Ruth Steinberger of Floyd, VA.

A work of this kind is very technical, and it is very easy for errors to slip by. In addition, the way houses are wired on the East Coast may not be the same on the West Coast. I would like to thank the technical readers, who helped minimize these problems: Bill Goode, the toughest electrical inspector that Roanoke County, VA, ever had; April Elkin, of Local #637, of Roanoke, VA; and Redwood Kardon, building inspector for the city of Oakland, CA.

Many companies helped with this book by providing photographic material and technical information about their products. I would like to give heartfelt thanks to Vivian Beaulieu of Marchant and Field; Ben Bird of Certified Insulated Products; Doug Kirk of King Technology, Siemens; Constance Malenfont of TayMac Corp.; Jean Miskimon of DeWalt Public Relations; M. F. Reed of Tytewadd Power Filters; and Jay Thomas of AFC Cable Systems. I would also like to thank the good people at Adjust-A-Box; Bussman; Greenlee Textron, Inc.; Hunter Fan, Hubbell, Ideal, Lutron, Magnus Industries, Inc., Technology Research Corp. and WhiteRock for their help.

CONTENTS

INTRODUCTION

My family has three generations of electricians—I am the third. As a kid, I remember working in spooky old buildings that had been around since the Civil War. The attics and basements of these houses were especially scary to me, but it was the crawlspaces I hated most. While lying in these damp or dust-choked caves, I remember pulling wires through floor joists spanned by spider webs clogged with insect carcases, attempting in vain to ignore the multilegged thing crawling up my leg—and all the while trying not to knock my head against another darn joist and hoping the flashlight wouldn't die. These are not fond memories, but I learned a lot. And it was literally from the ground up that I was taught about electricity and wiring a house. Now I'm both a master electrician and master plumber and have my own company.

I have written this book from lifelong experience and knowledge—some of which has been passed down through each generation. However, there is no one within my family to pass the gauntlet to—no fourth generation to pick up the trade. Therefore, by reading this book you will become heir to my knowledge and experience. You, in effect, will become the fourth generation.

Three main themes of the book are safety, design and materials. If you're doing the work yourself, I'll tell you how to do it safely. If you're not doing the work, you'll gain enough information from these pages to ask educated questions and to understand what the electrician is doing, and whether he or she is doing it correctly. You'll also learn what makes up a good electrical design and how to choose the best materials—not the lowest-cost materials—for your situation.

Safety is paramount

As an electrician, safety is a primary concern—for both my clients and me. Electricity can kill, so being alert while working with it and using safety equipment are both very important: I've received shocks before, and I would not be here today had it not been for a GFCI-protected receptacle. One of the aims of this book is to help you understand the basic principles of electricity so that you can give it the proper respect. You'll learn not only how electricity flows but also how to work according to a plan, so that any wiring job can be done safely and without fear. Throughout the book I'll offer safety tips that could save your life. Wiring can be tricky, so take your time and don't cut corners.

Tools are important to any person working in the building trades. Using the right tool for the right job will make the work go smoothly and safely. The same holds true for electrical tools; however, the wrong tool, or a low-quality tool, will not only cause headaches by making the job more difficult, but it could also seriously hurt or kill you. That's why I've devoted an entire chapter to tool use. In it I give you good background knowledge of electrical tools so that you will know what tools to buy and, even more important, which ones not to buy. I'll also illustrate how to use the tools correctly, not just in the tools chapter but throughout the book. And on p. 3 you'll find a key to help with the wiring diagrams in the book.

Meeting code is not good enough

Minimum code means exactly that—it's the absolute minimum required to pass inspection. If I had a house being built, I would want more in it than the absolute minimum. And I'm sad to admit that on today's jobs even minimum codes sometimes aren't enforced. Most inspectors are already backlogged and overworked—they have time only to check for obvious violations. They cannot trace every wire to be sure it goes to the right location, or even verify that the wire is the proper gauge. Therefore, you cannot assume the electrical system has been installed correctly or even safely just because it has been inspected and passed. Sadly, the bottom line is that it's normally up to the installer or homeowner to know what needs to be done and to see that it is done correctly. Knowing this, I try to give the reader the knowledge to know right from wrong, what works and what doesn't, and enough information so he or she can make intelligent decisions about the design of the electrical system.

You will learn how to develop a good, safe, high-quality electrical design, not one that simply meets minimum code. The electrical design is everything. One time I was called out to rewire a recently built house. It had already passed all electrical inspections, the walls were up and painted, and the owner had already moved in. The contractor had only been obligated to build to minimum specs: outlets were spaced a full 12 feet apart, with no receptacle outlet where it was needed for a specific piece of furniture, and a cheap, poorly designed electrical panel that was 99% full upon completion of the house was taking all the load it could handle, so nothing could be added (like a spa, for example), even if the panel had room for it.

The electrical system—in fact, the entire house—was built without any consideration as to what the owner needed. The owner had to pay twice: once for minimum code and the second time to get things custom-designed. A good design surpasses minimum code and takes the owner's needs into consideration.

Low-bid jobs are cheap—for a reason

It might be a surprise to some that it is impossible to obtain high-quality material on a low-bid job. Why? From the contractor's viewpoint, the object of the bid is to get the job. If I were to put together a bid that includes good-quality, high-end material, and my competitor makes a bid that includes cheap material, the bid will reflect this. My bid will be significantly higher than my competitor's, and I most likely will not get the job. For contractors, this book will illustrate where high-end material is appropriate and when you can get by with average-quality material (it's also possible to mix and match). With this knowledge, you'll be able to put together a bid that's reasonably priced, without compromising overall quality.

As a homeowner, you should know that if you choose the lowest bid possible, you may get exactly what you pay for. But if you specify in advance the type and grade of materials you want, so that all contractors are bidding with the same standards in mind, you can choose the lowest price with confidence that you haven't compromised quality. This book will give you enough knowledge to make informed decisions about the wiring system in your house, whether the house is new or old.

Simply tell all contractors bidding on the job the exact grade of material you want so that everyone is bidding on the same-quality items.

A book written from experience

This book is unlike any other wiring book on the market. Written by an electrician for homeowners, do-it-yourselfers and professionals alike, it is full of stories and experiences of exactly what happens when wiring a house. I even talk about some of the common mistakes that both pros (including myself) and do-it-yourselfers make so that you can avoid them from the outset.

I've always hated the standard how-to books on wiring because, for the most part, they're not written by practicing electricians. Instead, some desk jockey rehashes stuff from other books written by other desk jockeys. These books always pick perfect textbook situations with photos taken in a studio. They never tell you the problems you will encounter and what to do when things go wrong—or even the experiences of the authors. They can't, because the people writing the book have rarely done what they're telling you to do. I think both professionals and novices will appreciate my book because it's honest—I've done the work.

Key to drawings

In detailed drawings, wire labeling conventions are as follows:

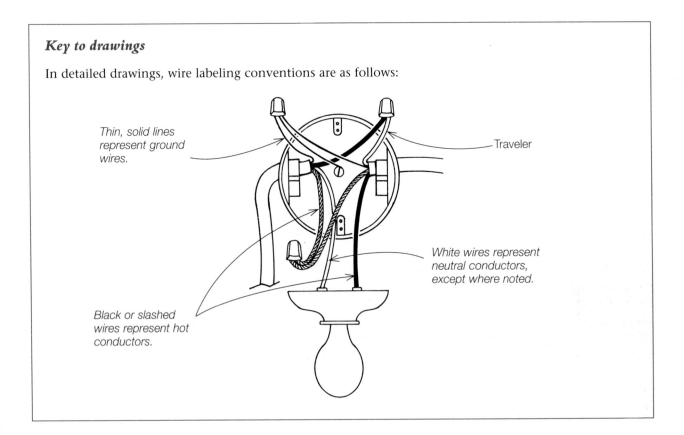

Thin, solid lines represent ground wires.

Traveler

White wires represent neutral conductors, except where noted.

Black or slashed wires represent hot conductors.

Chapter 1

RESIDENTIAL ELECTRICITY

I remember the first old house my Uncle Bud took me into when I was a child learning the trades back in the '50s. I even remember tripping on some funny-looking black wires in the attic and getting shocked as I grabbed onto a porcelain knob with a wire twisted around it. This was my introduction to old-house wiring. This type of old wiring is rather simple and quite unsafe by today's standards. Now, the amount of knowledge needed to make wiring safe is astronomical compared to the good-old days.

In this chapter I will help you understand the basics of electricity: how electrical systems work and how they evolved. You'll see all the different types of wires that you will encounter inside a house: what they look like and where they go. You'll actually look through the house walls and see parts of the house wiring system. And I'll help you understand some of the codes and standards that govern the products and labor that go into a house electrical system.

HOW ELECTRICITY FLOWS

To understand how electricity actually flows, visualize a wire like a garden hose, but instead of conveying water, it conveys electricity from one location to another. When you turn on a hose faucet, its flowing water pushes on water already in the hose, which pushes water out the other end. With electricity, an electron flows in one end of the wire, which knocks an electron out the other end. To get electricity started, you need electrical pressure, called *voltage*, provided either from the electrical utility or a battery. The higher the water pressure, the more water you get, and the higher the voltage, the more electricity you get. As water pushes through a hose, you have water flow, and as electricity pushes through a wire, you have *current flow*. The diameter of the hose can limit the amount of water you get out of it. What limits electrical flow is called *resistance*.

Elements of Electricity

Electricity is defined as the organized flow of electrons along a conductor. It is generated through heat, pressure, friction, light, chemical action or magnetism. The four basic elements of electricity are voltage, current, resistance and power. Ohm's law, a basic principle of electricity, states that voltage is directly proportional to the current and resistance. You can calculate any one value if the other two are known (E=IR, where E is the voltage, I is the current, and R is the resistance).

Voltage Voltage is the electrical pressure provided by a battery or other power source. The higher the voltage, the more current that the source can produce. In a basic formula without numbers, voltage is abbreviated by the letter E, meaning EMF or electro-motive force. Voltage is measured in volts, and a quantity of voltage is always followed by a capital V or the word volts (120V or 120 volts).

Voltage can be measured between two points on a circuit with a multimeter (a circuit is a complete path of current between the source and the load), or it can be calculated using Ohm's law if the current and resistance are known.

Current Electrical current is the organized flow of electrons from one point to another on a circuit. Electrons orbit around atoms just as the planets orbit the sun. When electrons get knocked out of orbit and start traveling down the circuit, knocking other electrons out of orbit, current is created. There are two types of current: alternating and direct.

Electricity that flows in one direction only is called *direct current (DC)* and can be seen occurring throughout the natural world (static electricity,

DC Circuit

DC circuits are simple. When closed, switch completes circuit, allowing current to flow to lamp to produce light, then back to positive terminal on battery.

lightning). Though DC generators do exist, DC is normally chemically generated, as in a battery. The current will flow like a river from its negative terminal to a load (a light or other appliance) and return to the positive terminal of the battery via another wire (see the drawing above).

Alternating current (AC) changes its direction of flow at regular intervals. AC, simply put, is generated by rotating a coil of wires through the magnetic lines of force—called flux—produced by a magnet, or vice versa. AC powers the receptacle outlets throughout a house. AC changes its flow along a sine curve at 60 complete cycles per second above and below a 0-volt reference point—one positive hump and then one negative hump equals one complete cycle (see the drawing on p. 6).

AC has a variety of voltages within its waveform, which are called instantaneous voltages. The point of maximum voltage is called the peak—it occurs at 90° on the curve. The effective voltage is 70.7% of peak voltage and occurs at 45° on the curve. The effective

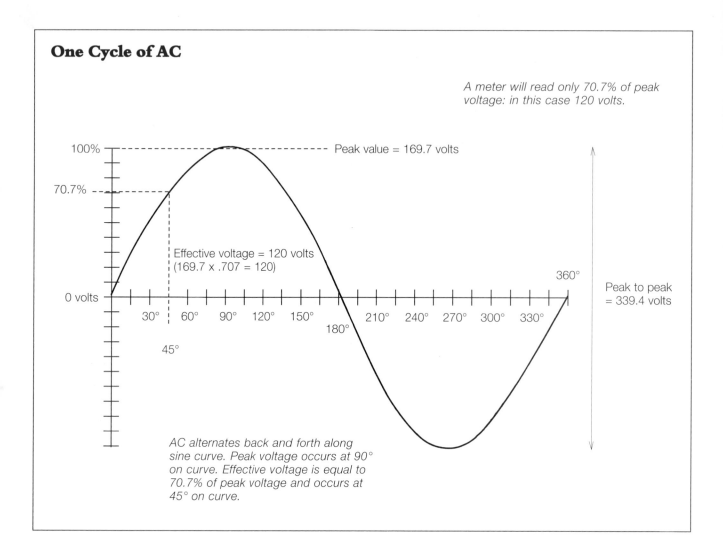

One Cycle of AC

A meter will read only 70.7% of peak voltage: in this case 120 volts.

100% — Peak value = 169.7 volts

70.7%

Effective voltage = 120 volts
(169.7 x .707 = 120)

360°

0 volts

Peak to peak
= 339.4 volts

30° 60° 90° 120° 150°   210° 240° 270° 300° 330°
180°

45°

AC alternates back and forth along sine curve. Peak voltage occurs at 90° on curve. Effective voltage is equal to 70.7% of peak voltage and occurs at 45° on curve.

voltage has the same heating effect as a DC circuit of the same voltage—effective AC volts will provide the same power as DC volts of the same rating: For example, 120 AC volts is the same as 120 DC volts.

Current can be increased by raising the voltage or lowering the resistance. In a basic formula without any numbers, current is always represented by the letter I. It is measured in amperes (amps, for short), and the amperage will follow any numerical amount (e.g., 20 amps).

Electrical current can be measured by inserting an ammeter in series with the wiring (cutting the wire and attaching a meter lead to each wire end), but it

is much safer to use a clamp-on meter (see p. 43). This type of meter clamps around the wire itself and measures the electromagnetic waves emitting from the wire to determine the current. The latter method is much safer than cutting the wires and is universally used by electricians. If the voltage and resistance are known, current can be calculated using Ohm's law: I=E/R.

Resistance As current flows through a material, it meets the inherent resistance of that material—the physical opposition to current flow (like a dam trying to hold a river back). The opposition is simply electrons refusing to be stripped of their atoms and be sent down the wire. The higher the material's resistance, the more energy it will take to strip the electrons off. And the higher the resistance, the less current that can flow. The only way to increase the current in a circuit with a specific resistance is to raise the voltage, which increases the pressure to compensate for the fixed resistance.

In a basic formula without any numbers, resistance is represented by the letter R. It is measured in ohms, and a number quantity is always represented with the Greek letter omega (Ω) adjacent to the numerical amount of resistance, or ohms written out (30Ω or 30 ohms). Resistance is measured by a volt-ohm meter (VOM) or a multimeter, or it can be calculated if the voltage and current are known: R=E/I.

Resistance can also be the property of an appliance that converts electrical energy to heat energy. To see how, let's take a common 240-volt, 4,500-watt water heater that's pulling 18.75 amps. As the 18.75 amps flow into the water-heater element, the element produces heat as it opposes the electron flow (the element has a resistance of 13 ohms). This opposition produces heat—exactly 4,500 watts worth—which turns the cold water to hot.

Power Current and voltage provide power, expressed as true power or apparent power. ***True power*** is the power actually used by the load and is expressed in watts. For DC circuits or AC circuits that are purely resistive, true power can be calculated using the formula P=EI. ***Apparent power*** is measured in volt-amps (VA). In some AC circuits, because of capacitance and inductance, the voltage and current will behave differently than in resistive circuits. When such voltages and currents are used in power equations, the result is apparent power. Apparent power = watts/power factor. VA is always greater than watts.

Common VOMs and multimeters cannot measure wattage (although commercial units can), but it can be calculated if two of the following are known: voltage, current and amps.

1. If only voltage and current are known, use P (watts)=EI. For example, if E equals 120 volts, and I is 10 amps, the power has to be 1,200 watts (120 x 10).

2. If current and resistance are known, use $P=I^2R$. For example, if I equals 10 amps, and R is 12 ohms, the power has to be 1,200 watts (10^2 x 12).

3. If only voltage and resistance are known, use $P=E^2/R$. For example, if E is 120 volts, and R is 12 ohms, the power has to be 1,200 watts ($120^2/12$).

WIRES: PAST AND PRESENT

Wire provides a means of conveying a quantity of amps from the voltage source to a place where you want to use it. Remember the analogy between a water hose and electrical flow? Well, I talked about how the diameter of the hose can limit the water flow: A very small hose will deliver water at a slow rate, while a large hose will deliver water at a faster rate. It works the same way with wiring.

Is It Wire or Cable?

The terms wire and cable are used interchangeably by most electricians. Technically, a wire is an individual conductor, and a cable is composed of two or more wires surrounded by a jacket. However, most electricians will still refer to cable simply as wire. When an electrician requests 12-gauge wire for a particular spot, he normally wants 12-gauge cable, unless he were running individual wires within a conduit system. On the other hand, large-diameter wires are often referred to as cables. And the word wire can be both singular and plural. Because this gets very confusing, for clarity, it's better to say "conductor" as opposed to saying "wire"—a conductor, without debate, is a single wire.

Wire Gauge

Just as the diameter of a hose is important to delivering water, the diameter of a wire is important to delivering electrical current. Resistance provides the opposition to current flow in a wire: the more resistance, the less current you get. A large-diameter wire delivers a larger flow of current compared to a small-diameter wire because it has less resistance.

The reference to wire size in the United States is called the American Wire Gauge (AWG) and is represented by gauge numbers. The smaller the gauge number, the larger the wire diameter (see the drawing below). For example, a 14-gauge wire is physically smaller in diameter than a 10-gauge wire. A wire with a large diameter will have less resistance to electrical current and will therefore be able to carry more current safely. *Too much current flowing through a small-diameter wire will overheat the wire, damage its insulation and could start a fire.* On the other hand, always installing large-diameter wire is cost-prohibitive and might damage receptacles and other items that aren't made to take the increased diameter.

Tiny wires, each the size of small string, are numbered 18 and 16 gauge and are good only for low-voltage equipment like doorbells, fixture wiring and cords. Wires numbered 14 and 12 gauge are for receptacle and switch circuits and small appliances. Wires numbered 10, 8 and 6 gauge are for appliances that need a lot of current. And the smallest numbers, those even smaller than 0 (pronounced "ought"), like 0000 (four ought, or 4/0), are for the heaviest current users of all: the connections to the utility. (Some references will refer to ought as "naught.") Larger than 4/0, the wire is described in its cross-sectional area of circular mils: kcmils (a mil is $\frac{1}{1,000}$ of an inch). But it would be rare to use wire larger than 4/0 in a house. The chart on the facing page illustrates the service-entrance ratings of various service-entrance (SE) conductors—how much current can be run through a certain gauge SE wire. This chart will come in handy when you're sizing your electrical service.

So just how do you tell different gauges apart? It's hard. Even professionals, like me, have a hard time telling different-gauge wires apart. You can buy a

Cross Sections of Copper Conductors

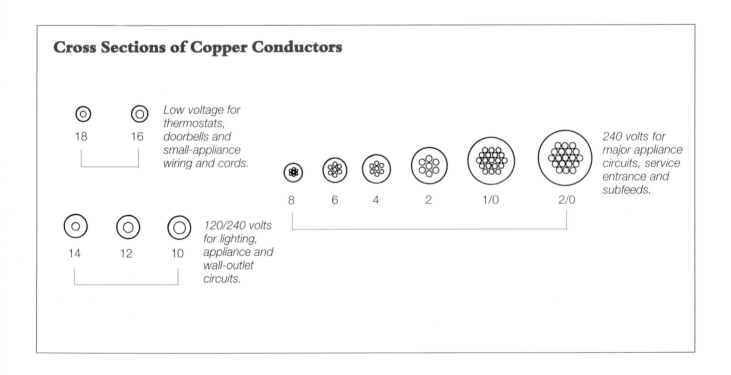

18 16   Low voltage for thermostats, doorbells and small-appliance wiring and cords.

8   6   4   2   1/0   2/0   240 volts for major appliance circuits, service entrance and subfeeds.

14 12 10   120/240 volts for lighting, appliance and wall-outlet circuits.

wire gauge, but they're hard to find. Of course, the wire manufacturers will tell you that the wire gauge is stamped on the cable jacket. But every electrician knows the folly of this. On common 14-, 12- and 10-gauge cable, manufacturers do stamp the gauge size on the cable jacket. But some manufacturers don't use paint or ink; they only make a slight impression on the sheath, and it is very hard to read. Some wire is so small in diameter that letters painted on the jacket are only about $\frac{1}{32}$ in. to $\frac{1}{16}$ in. tall. You'll need a microscope to read lettering that size. Other times, the part that has the gauge written on it has been cut away. The bottom line is, until you get enough experience to tell at a glance the gauge of a wire, or locate a wire gauge, it's best to have wire samples on hand for direct comparison.

Copper or Aluminum Wiring

There are two standard wire materials for residential wiring: aluminum and copper. Normally, the wiring for receptacles and switches within a residence is copper—and for good reason. Small aluminum wires have to be handled very carefully because they break easily. And they must be installed and spliced only with code-approved material identified for such, and sometimes that material is hard to find. For the larger-diameter wires (6 gauge and up), electricians tend to use both copper and aluminum. Aluminum has a slight cost advantage over copper, which is more apparent in the larger-diameter wires.

When exposed to air, bare aluminum will oxidize, which will insulate the strands of aluminum from each other, as well as from the connector itself, increasing the resistance and degrading the metal. Therefore, whenever aluminum has its insulation stripped off, an anticorrosion coating must be put on *all* the bare wire to prevent oxidation—not just on the part connecting to the terminal. Some aluminum manufacturers recommend wire-brushing the aluminum to remove any oxide present on the surface before applying the antioxidation compound.

Service-Entrance Conductors

Copper	Aluminum and copper-clad aluminum	Service rating in amps
AWG	AWG	
4	2	100
1	2/0	150
2/0	4/0	200

200-amp service is the most common in a house. (Table is partially reproduced from p. 178 of the NEC.)

The Best Conductors

Copper and aluminum are used as common conductor materials because of their physical makeups (the atomic structure of the material). In addition, the rarity of the metal and the expense of manufacturing it into a usable form must be considered. Most metals conduct electricity, but some are better conductors than others. The Periodic Table of Elements lists the electrical conductivity of the elements—the higher the number, the better the conductor. In reference to the better metals used as conductors, they are listed in order of decreasing conductivity as follows: silver at .63, copper at .59, gold at .45 and aluminum at .377. Silver is obviously the best conductor but is much too expensive to be used as a common electrical wire, just as gold is. However, gold is used as a coating on some stereo patch cords, not because it is such a great conductor but because it doesn't tarnish or corrode. Practically, this leaves copper and aluminum as the primary material for wiring.

Conductor Resistance at 75°C (167°F)

Size AWG/kcmil	Copper ohm/1,000 ft.	Aluminum ohm/1,000 ft.
14	3.07	5.06
12	1.93	3.18
10	1.21	2
8	0.764	1.26
6	0.491	0.808
4	0.308	0.508
3	0.245	0.403
2	0.194	0.391
1	0.154	0.253
1/0	0.122	0.201
2/0	0.0967	0.159
3/0	0.0766	0.126
4/0	0.0608	0.100

Choose the Correct Wire Gauge

When making a decision about which wire gauge to use and whether to use copper or aluminum, it's helpful to know which conductor has the lowest and highest resistance. This will help you determine the voltage drop that would occur across the wire as the current flows through it. If the wire is taking away a substantial amount of voltage from the load, the load will operate less efficiently—and in worse-case conditions, burn out. You want to use material that will keep the voltage drop across the wire down to about 3% to 5% of the input voltage.

For example, let's assume you want a lamp pulling 5 amps at the end of the driveway. You'll need about 1,000 ft. of wire to and from the panel (500 ft. to the lamp, 500 back). What gauge wire should you use? Let's calculate the voltage drop to see if you can get by using the less-expensive 14-gauge wire. Using the chart above, you can see that 14-gauge copper wire has 3.07 ohms resistance for the 1,000 ft. Five amps being pulled through this resistance will develop 15.35 volts across it (E = IR, 15.35 volts = 5 amps x 3.07 ohms). In a standard 120-volt circuit, this leaves only 104.65 volts for the load, which means you'd be losing 12.8% of the voltage to the resistance. This situation is unacceptable.

Let's try 12-gauge copper wire. Its resistance decreases to 1.93 ohms for the same 1,000 ft. Sending 5 amps through it will reduce the voltage drop to 9.65 volts (5 amps x 1.93 ohms) across the wiring, meaning that you'd be losing 8% to the wiring. A dramatic difference but still too high.

Now let's try 10-gauge copper wire. Its resistance is 1.21 ohms for 1,000 ft., which results in a voltage drop of 6 volts. A voltage loss in the area of 5% is acceptable. Therefore, to wire the lamp, you'll need 10-gauge wiring.

Knob and Tube

The old wiring in the turn-of-the-century house where I was shocked as a child is called knob and tube, and it was used in the late 1800s and the early 1900s. I'm sorry to say that many old houses are still wired and operating with it today. This wire may have been considered safe at the time it was originally used, but it is not adequate to keep up with modern demands. The wiring gets its name from the round porcelain brackets (knobs) that hold the individual wires (this is what I grabbed onto when I was shocked). To run wire, the electrician had to drill through wood beams and then insert clay (and later porcelain) tubes through which the wires would pass (see photos on the facing page). Throughout a typical old house, you'd see two large-diameter black wires traveling adjacent to each other from one porcelain knob to another: one hot and one return (no ground). The wires were copper and were covered with a dark-colored rubber with either a single or double cotton saturated braid.

Many buildings utilizing the knob-and-tube system were powered by rechargeable DC batteries. My Uncle Bud still remembers being in church as a child when the lights would get dimmer and dimmer as the old Delco plant lost its power. The church-goers would strain their eyes reading the Bible until they thought they were going blind. The rechargeable systems finally made way for the utility company.

Round Solid Porcelain Knobs

63N3925—Old Code No. 5½ solid porcelain knob. Height, 1 9/16 inches. Diameter, 1 inch. Hole, ¼ inch. Groove, 5/16 inch. Weight, per 100, 8½ pounds. Price, each, **1c**; per 100, **92c**; per 1,000........................**$9.00**

63N3927—New Code No. 5½ solid porcelain knob. Height, 1 9/16 inches. Diameter, 1⅛ inches. Hole, ¼ inch. Groove, 5/16 inch. Weight, per 100, 11½ pounds. Price, each, **1½c**; per 100, **$1.18**; per 1,000..**$11.20**

63N3929—No. 4 solid porcelain knob. Height, 1 11/16 inches. Diameter, 1½ inches. Hole, ⅜ inch. Groove, ⅜ inch. Weight, per 100, 20 pounds. Price, each, **2c**; per 100, **$1.57**; per 1,000....**$15.10**

63N3931—No. 4½ solid porcelain knob. Height, 1⅞ inches. Diameter, 1½ inches. Hole, ⅜ inch. Groove, 7/16 inch. Weight, per 100, 21½ pounds. Price, each, **2¼c**; per 100, **$1.76**; per 1,000....**$17.00**

Round Split Porcelain Knobs

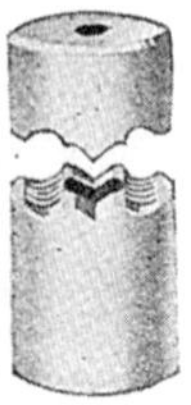

63N3935—Old Code No. 5½ porcelain split knob. Height, 1¾ inches. Diameter, 1 inch. Hole, ¼ inch. Grooved to take two No. 12 or 14 wires. Weight, per 100, 9½ pounds. Price, each, **2c**; per 100, **$1.56**; per 1,000 **$15.00**

63N3937—New Code No. 5½ porcelain split knob. Height, 1¾ inches. Diameter, 1⅛ inches. Hole, ¼ inch. Weight, per 100, 11 pounds. Price, each, **2¼c**; per 100, **$1.75**; per 1,000.**$16.90**

Porcelain Tubes

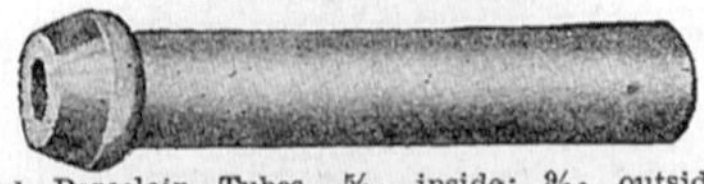

Unglazed Porcelain Tubes, 5/16 inside; 9/16 outside. Take either 14, 12 or 10 single braid rubber-covered and weather-proof wire. Required wherever a wire is drawn through a partition or joist of any kind. Length given is from under-head to end.

Article Number	Length, Inches	Each	Per 100	Per 1,000	Weight per 100 lbs.
63N3902	3	1c	$0.78	$ 7.15	7½
63N3904	4	1½c	1.25	11.95	9
63N3906	6	2c	1.73	16.80	13
63N3908	8	4c	3.30	31.35	15½

Porcelain Cleats

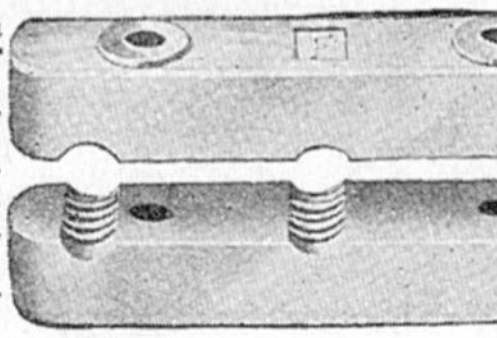

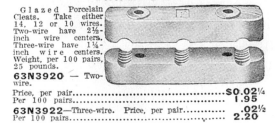

Glazed Porcelain Cleats. Take either 14, 12 or 10 wires. Two-wire have 2½-inch wire centers. Three-wire have 1¼-inch wire centers. Weight, per 100 pairs, 25 pounds.

63N3920 — Two-wire.
Price, per pair..**$0.02¼**
Per 100 pairs..**1.95**

63N3922—Three-wire. Price, per pair............ **.02½**
Per 100 pairs.. **2.20**

New Code, Rubber-Covered Copper Wire, Single Braid

Solid conductor wire, insulated with rubber compound over which is one cotton saturated braid. Recommended for any open wiring on cleats, tubes or knobs, and for loom and moulding wiring.

Each unbroken coil bears Underwriters' Inspection tag. Full coils contain 100 feet or 500 feet, or we will cut wire to any desired length.

Article Number	Size	Weight per 100 ft Pounds	Per Foot	Per 100 Feet	Per 1000 Feet
63N3015	14	3	1½c	$1.40	$10.25
63N3020	12	4½	2½c	2.20	19.75
63N3025	10	6	3¼c	3.00	27.40

New Code, Rubber-Covered Wire, Double Braid

Offers better protection than single braid wire. Should be used in all metallic conduits, flexible or rigid. Has one solid conductor, insulated with rubber compound over which are two saturated cotton braids.

Article Number	Size	Weight per 100 ft. lbs.	Per Foot	Per 100 Feet	Per 1000 Feet
63N3040	14	4	2c	$1.70	$16.20
63N3045	12	5½	2¾c	2.35	22.25
63N3050	10	7	3½c	3.20	29.90
63N3055	8	10	5c	4.55	43.95
63N3060	6	13	7½c	7.05	67.50
63N3065	4	18½	10½c	9.90	96.25

Duplex New Code Rubber-Covered Wire

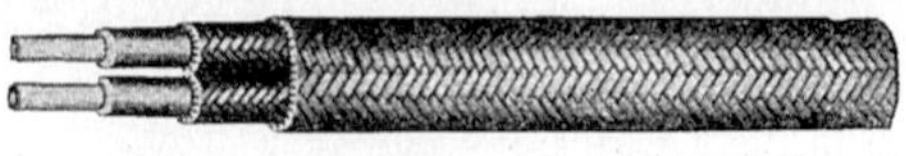

Consists of two solid conductors, each insulated with rubber compound, over which is one saturated cotton braid. Conductors so insulated are laid parallel and covered over all with saturated cotton braid. Convenient for wiring in metallic and non-metallic conduits.

Article Number	Size	Per Foot	Per 100 Feet	Per 1000 Feet
63N3080	14	3¼c	$3.00	$27.50
63N3085	12	5c	4.75	44.95
63N3090	10	7c	6.60	65.25

New Code Electric Fixture Wire

For wiring Electric Chandeliers, fixtures, brackets, lamps, etc. Also used for short lighting circuits. Solid copper conductors. No. 18 size, has 1/64-inch, new code rubber compound covering. No. 16 size, 1/32-inch new code rubber compound covering. Each is covered with saturated cotton braiding.

Article Number	Size	Per Foot	Per 100 Feet	Per 1000 Feet
63N3102	18	1c	$0.83	$ 7.70
63N3107	16	1½c	1.20	10.45

Weather-Proof Copper Wire

For use on outside work only. Not to be used indoors. Withstands weather better than rubber-covered wire. Conductor is of solid copper wire, covered with a triple braid of weather-proof insulating material.

Article Number	Size	Wt. per 100 ft. Pounds	Per Foot	Per 100 Feet	Per 1000 Feet
63N3125	14	2½	1½c	$1.15	$ 9.95
63N3130	12	3½	1¾c	1.55	14.50
63N3135	10	5¼	2½c	2.30	21.25
63N3140	8	7½	3½c	3.15	29.25
63N3145	6	11	5c	4.70	43.65
63N3150	4	16½	7½c	6.90	63.90

These photos—one from an early Montgomery Ward catalog and one from a Sears & Roebuck catalog—show wire and cable that was available around 1900-1920. The knobs, tubes and cleats were used for knob-and-tube wiring systems. (Photos by Scott Phillips)

However, the old knob-and-tube wiring never changed. Over the years it performed admirably. However, as homes modernized, more and more requirements were placed on the wiring. At one time it used to be considered modern just to have one light and one receptacle per room; now it means nothing to have a half-dozen outlets per room with appliances in every corner. The knob-and-tube wiring just couldn't take it. To add to its problems, the rubber insulation on the wires would eventually rot and break away, leaving patches of bare conductor. Modern wiring is different from that of yesteryear. Thermoplastic has replaced saturated cotton as a jacket material; flame-retardant, heat-resistant thermoplastics have replaced rubber as the wire insulation; and lead is no longer used.

Flexible Armored Cable (AC)

AC (armor-clad) cable came into being as an offshoot of the early conduit systems. Early wiring systems used old abandoned gas pipes as conduit to hold and protect the wiring. Where no gas pipes were present, new conduit had to be run. Short conduit runs proved to be expensive to install, so cable manufacturers figured out a way to build the conduit around the wire at the factory. And flexible AC cable was born.

BX is a common name given to this type of cable. I do not know the origin of the term BX, but at least one source indicates that it was derived from an abbreviation of the word Bronx, where the cable was once made. BX is a registered trademark of General Electric Company. However, it has become a generic term to all electricians.

The early BX system consisted of two or three insulated copper conductors (no separate ground wire), from 14 gauge to 10 gauge (currently available up to 1 gauge), installed inside a flexible, spiraled steel armor, which if needed, served as a ground. The system was called BX, not Greenfield as I have read in other sources. Greenfield uses the same flexible steel jacket as BX, but the wiring is inserted into the jacket at the job site, not at the factory. BX was an improvement over knob and tube because the flexible metal armor protected the wiring and made it easier and faster to install.

Today's BX, also called AC, has both steel and aluminum armor with thermoplastic insulated conductors individually wrapped with a waxed paper jute. It is available in one-, two-, three- and four-conductor assemblies. It also has a 16-gauge aluminum bonding wire within the jacket but is not meant to be wired as an equipment grounding conductor (see the top drawing on the facing page). It makes contact with the BX metal for its entire length and is part of the grounding system in that it helps reduce the grounding resistance of the flexible armor. It is not to be wired to anything. It can just be cut off or used to help hold the plastic antishort bushing that goes into the end of the cut armor. Grounding, as before, is done via the armor itself.

Armored cable is rated according to the characteristics of the insulation surrounding the wire: Type ACT has a PVC insulation—good up to 60°C in dry locations. If an H is added, the temperature characteristics of the cable's insulation have been increased. For example, cable type ACTH is good up to 75°C, and type ACTHH up to 90°C. Today, most AC cable is rated up to 90°C because of its THHN conductors (I'll talk more about cable insulation and its importance on pp. 24-26). Armored cable is covered in Article 333 of the NEC.

Metal–Clad Cable (MC)

MC (metal-clad) cable is a direct offshoot of AC cable. AC was designed from early on as small wire sizes carrying no more than 30 amps of current. When the need for heavier-gauge conductors that could carry more current became evident, MC cable was born. MC cable has either interlocking armor, like AC, or a smooth or corrugated tube armor (see the middle drawing on the facing page). The conductors are not individually wrapped as in AC cable; rather, the entire bundle is wrapped in paper or plastic. MC cable normally comes in gauges of 6 and higher and with either three or four insulated conductors and a bare grounding wire. However, at least one manufacturer extends the range down to 18 gauge with multiple conductors.

Nonmetallic Cable (NM)

NM cable consists of two or more insulated conductors in a single nonmetallic jacket (two insulated conductors in an NM jacket is called a duplex). It is not a modern invention. Duplex cable has been around since before 1918. Early duplex NM cable consisted of two insulated conductors in a silver-colored jacket or a braided, brown-jacketed cable. This silver-colored jacket cable was used to wire houses in the early 1950s; and the brown-jacketed cable was used to wire houses in my section of Virginia in the 1920s and 1930s.

Modern NM cable (also called Romex, a registered trademark of the General Cable Company) came into use around 1965. There are four basic types of NM cable—old style NM, NMC, NMB and UF—and each has a specific job. Old-style NM was the most common cable used in home wiring until a few years ago. It is now replaced by NMB cable because of the need for a higher insulation temperature. Old-style NM is rated at 60°C, and NMB is rated at 90°C. Generally, NM-type cable has two or more thermoplastic insulated wires, one bare grounding wire, a paper insulation surrounding the conductors and a thermoplastic jacket, or sheath, over that (see the bottom drawing at right). NMB cable is the NM style that is currently used around the home for receptacles, lighting and small-appliance circuits. Modern NMB cable can be used for both exposed (as long as it is in no physical danger) and concealed work in dry locations, but it cannot be embedded in masonry, concrete, adobe, dirt or plaster. NMC cable is the same as NMB cable, but its outer jacket is corrosion-resistant.

Type UF (underground feeder) cable looks like NM cable, but UF has the grounding wire, the neutral wire and the hot wire embedded in solid thermoplastic. UF cable is used in wet locations, such as an underground circuit to an outdoor light, where NM cable would not be allowed by code.

NM cable is normally sold in 250-ft. rolls. Some stores that sell to do-it-yourselfers sell the cable in smaller lengths; however, the markup may be considerable.

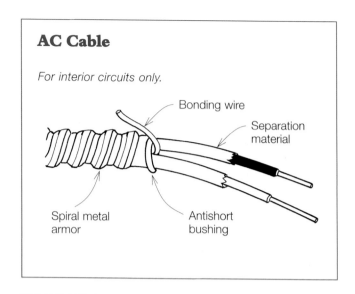

AC Cable

For interior circuits only.

Bonding wire
Separation material
Spiral metal armor
Antishort bushing

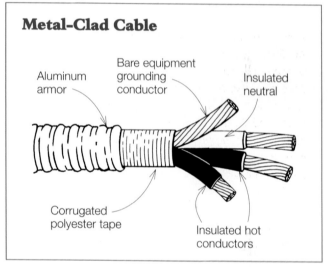

Metal-Clad Cable

Aluminum armor
Bare equipment grounding conductor
Insulated neutral
Corrugated polyester tape
Insulated hot conductors

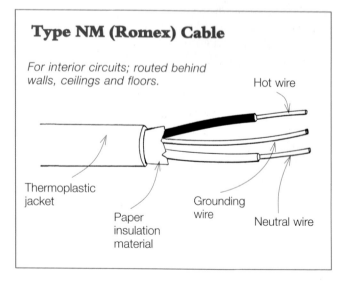

Type NM (Romex) Cable

For interior circuits; routed behind walls, ceilings and floors.

Hot wire
Thermoplastic jacket
Paper insulation material
Grounding wire
Neutral wire

What 'Neutral' Means

The neutral conductor is the conductor that eventually leads back to the center tap of the utility transformer. It is connected to earth at two locations: first at the utility transformer, and second at the main panel. Therefore, it is referred to as the grounded conductor. A grounding conductor makes the physical connection to the ground rod. However, no neutral current flows into the earth via the grounding conductor.

Neutral current is commonly referred to as return current because once it flows through the load, it will flow back through the household wiring (white insulated wires), back through the main panel, back through the service entrance (normally the braided neutral), back through the utility neutral (the bare support conductor that the two insulated conductors hang from), back into the transformer center tap and back into the transformer windings where it originated. All electrical current must make a complete loop. The current starts at the utility transformer and must therefore return to the transformer; the neutral is simply the return part of that loop.

Service-Entrance Cable

Type SE (service entrance) cable brings power into the house from the overhead splice at the utility cables. SE cable comes in different styles: U, R and USE. Style U is flat (see the top photo on the facing page), approved only for above-ground use. This is the cable you see attached to the outside of most houses. It has two black insulated wires (sometimes one has a stripe for phase identification) surrounded by many strands of braided bare wire for the neutral (for a related discussion, see above), covered by a layer of glass-reinforced tape and protected by a PVC jacket. The jacket is flame-retardant, moisture-resistant and almost sunlight-resistant.

Style R is round and has three insulated conductors with one bare grounding wire (see the middle photo on the facing page) covered with a layer of glass-reinforced tape and protected by a PVC jacket. Like style U, R is only approved for above-ground use.

Individual insulated conductors rated USE (underground service entrance) are normally used when the service must be located underground or within conduit (see the bottom photo on the facing page). SEU (flat), SER (round) and USE (underground) are available in gauges of 6 and higher.

SE cable can also be used to provide power to some appliances, such as 240-volt only heat pumps and electric furnaces. However, it cannot be used where an insulated neutral is required. SEU cable cannot be used to feed 120/240 appliances, such as electric dryers and stoves. Such appliances need three insulated conductors plus ground (two insulated conductors for the 240-volt load, and a return neutral for the 120-volt load), and standard SE cable has but two. However, SER cable can be used to feed 120/240 volt loads because it has three insulated conductors with an additional bare grounding conductor. SE cable is covered in detail in article 338 of the NEC.

ALCAN S STABILOY® AA-8000 AL TYPE SE CABLE STYLE U XHHW-2 600V 3 CDRS 1/0 (UL) 1995

ALCAN S STABILOY® AA-8000 AL TYPE SE CABLE STYLE R XHHW-2 600V 3 CDRS 1/0 1 CDR 2 (UL) 1995

ALCAN S 4/0 COMPACT STABILOY® AA-8000 AL XLPE 600V USE-2/RHH/RHW-2 (UL) 1995

U-style SE cable has two insulated conductors (top). One is black, and the other is black with a red stripe for phase identification. The neutral has many strands braided around the two insulated conductors. Surrounding all this is glass-reinforced tape and a PVC jacket. (Photos this page courtesy of Alcan Cable)

Style-R SE cable is round and contains three insulated conductors plus a grounding wire (middle). It can only be used above ground. The conductors are surrounded by glass-reinforced tape and a PVC jacket.

Style-USE cable has individual conductors that can be buried or placed within conduit. The conductors are insulated with cross-linked polyethylene (bottom).

POWER GENERATION

Electricity has been around since the beginning of time. To create electricity on demand, as opposed to watching it work in nature, has always been a most challenging problem. Most societies tried to use chemicals and get them to react with other substances to produce a current flow of long enough duration to be useful. The most common example of this is a battery. Direct current (DC) is created through the chemical reaction of materials inside a battery. The current leaves the negative side of the battery, goes through the load, such as a flashlight bulb, and back into the positive side of the battery. A complete circuit has to be made from the negative terminal to the positive terminal of the battery for the current to flow.

Codes and Standards

There are a few organizations in the United States that provide help in creating and understanding electrical standards and materials. Each of the organizations has direct input in the installation, manufacture, testing or application of electrical material. Without these organizations, there would be chaos within the industry. For example, a typical receptacle is built pretty much the same way according to set standards. Without standards, one manufacturer's receptacle design would be different from another; the ground slot might be on the side, and the wide slot might be hot instead of neutral.

The American National Standards Institute
The American National Standards Institute, Inc. (ANSI), is a coordinating organization for voluntary standardization in the United States. ANSI coordinates the efforts and results of other standard-making organizations. If a standard is agreed upon, it is given a title and designation number known as an American National Standard.

The National Electrical Code
The National Electrical Code (NEC) is the electrician's bible. The code was first conceived in 1881, in Richmond, Virginia, at a meeting of the National Association of Fire Engineers. This meeting resulted in a published electrical code in 1895. In 1897 it became the National Electrical Code.

The NEC is published by the National Fire Protection Association (NFPA). The NEC sets voluntary nationwide standards, which are adopted by official agencies to keep electrical wiring and installation of wiring standardized and safe. For example, the NEC standardized the color-coding of wires: It's nice to see a white wire and know that it is a neutral and that the black wire is hot.

Although the NEC is supposed to be consistent across county and state lines, you may find differences from locality to locality. Local inspectors interpret the code differently—I literally have to wire differently in every county—but the differences are minor. Without the NEC the differences would

---

The problem with developing electrical current through chemicals is that its source eventually dies. Batteries producing DC eventually produce less current as time increases. Once the chemical process is completed, we call the battery "dead." DC generators can mechanically produce DC as long as power is available to turn the generator. However, there was no practical method to transmit the DC power over long distances without a significant loss of voltage.

AC, on the other hand, could be transmitted over long distances—without loss of voltage—with the invention of the transformer (more on transformers later). An AC generator moves coils of wire around a magnet. Every time the coils of wire move, or cut, through the magnetic flux, enough energy is transferred from the field and into the moving coils to produce current flow within the generator. The problem has always been keeping the coils or magnet constantly moving—it takes energy to produce energy. On a small scale this is not a problem: A man pedaling a bicycle with the rear wheel replaced by an AC generator will work. On a large scale, however, you need a lot of power to generate enough energy to supply thousands of homes and businesses. These days, large AC generators produce vast amounts of electricity. These huge generators run on nuclear or hydroelectric power, fossil fuels, as well as solar power to produce massive amounts of electricity.

be major to the extent that there would be total chaos.

The code is revised every three years, which is how electricians know their age—by how many old code books they have. Revisions are compiled and reviewed by the different code committees that write the different sections of the code.

The National Electrical Manufacturers Association
Founded in 1926, the National Electrical Manufacturers Association (NEMA) is the largest trade association in the United States that represents the electrical industry. NEMA establishes voluntary standards in the electrical manufacturing industry to eliminate problems between designer, manufacturer and purchaser. If all equipment is built to NEMA standards, working with the equipment and ordering replacement parts for that equipment is easy.

Underwriters Laboratories
Underwriters Laboratories (UL), Inc., is an independent testing laboratory that sets standards for products and certifies that those products comply with the standards. In simple terms, UL verifies that a manufacturer's product will do what the manufacturer says it will. (The UL is not the only testing lab, however; it is just the best-known.)

The UL standards come in direct contact with the electrician and homeowner in residential construction. Many appliances or equipment on the electrical circuit require a label indicating that the UL has tested the equipment for its intended use, or it may not pass inspection. Throughout the NEC you will see the word "listed." Listed means that the equipment to be used for a specific application must be approved by a third party, such as UL, or by the equipment manufacturer for that application. For example: If you install a fan in the shower area itself, make sure the fan is UL listed (or listed by another recognized testing agency) for that location. In this case the fan would be labeled for wet locations and would need ground-fault protection, or it will not pass the rough-in inspection.

Long-Distance Power Transmission

Transmitting voltage and current from one location to another—without voltage loss—was solved by the invention of the transformer. A transformer is nothing more than two or more tight coils of wire placed next to each other. When current flows through a wire, magnetic lines of force (magnetic flux) emanate from that wire, just as they emanate from a magnet. The magnetic flux created by current flowing through the first coil (called a primary) will induce a voltage in the second (called a secondary) that is proportional to the number of wires (windings) within each coil. The voltage from one coil to the other is proportional to the number of windings in each coil. For example, if the primary coil has fewer windings than the secondary coil, the voltage would be proportionally greater on the secondary—only the current would be smaller (see the drawing on p. 18). There are two types of transformers: step-up and step-down. The step-up transformer increases voltage from the primary coil to the secondary, but lowers the current. The step-down transformer reduces voltage from the primary coil to the secondary, but increases current.

Here's how the system works: The step-up transformer at the power plant generates an extremely high voltage (pressure) with a low current, which flows through the high-power transmission lines you see on the tall steel towers across the

Transformers

Step up

Step-up transformer increases voltage but decreases current.

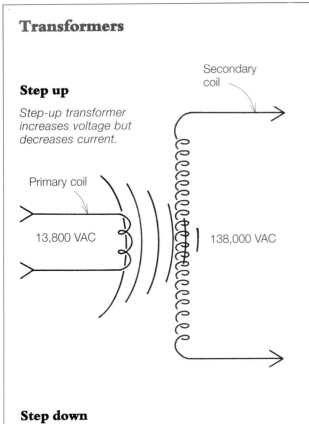

Secondary coil

Primary coil

13,800 VAC

138,000 VAC

Step down

Step-down transformer decreases voltage but increases current.

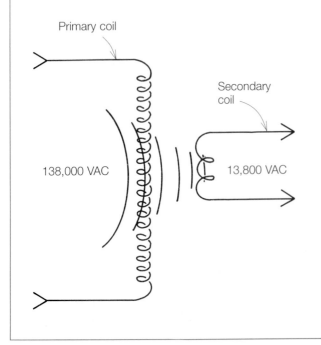

Primary coil

Secondary coil

138,000 VAC

13,800 VAC

country. These power lines connect to a step-down transformer at a local substation. This transformation occurs many times throughout the voltage distribution path across the country and local areas (see photos on the facing page).

The power leaves the substation on three hot power lines and travels up and down the roads and highways (there may be significant regional differences across the country). Power is distributed locally in three phases or power sources. All three phases share one neutral and have the same voltage, but they are out of phase with each other. This simply means that the one voltage was started in time a little before another. Because this phase difference is only useful to heavy industry, it makes no difference which phase powers the residence. A residence simply sees the three phases as three possible power sources to pick from, and the utility company will pick one for you. Eventually, the lines become separated as they feed different areas until a single line is left to feed the residences. The step-down transformer on the pole outside your house converts the utility voltage from several thousand volts down to 120/240 volts for use in your house.

Power from Utility Pole to House

Let's take a close look at the utility power line right outside your house. There are too many utility wiring situations to describe what actually occurs on every home utility pole, but I'll discuss the most common situation: one utility line (wire) at the top of the pole and another a few feet below it (see the drawing on p. 20). The top wire is hot and carries several thousand volts. The lower wire is the neutral/ground, or common, which is connected to the earth via ground rods at certain poles up and down the line. (Again, there are significant regional differences. For example, in California, the neutral wire is not carried anywhere near the hot wire.) The transformer at the house taps into the hot and neutral lines and steps down the voltage to 120/240 volts, which become available on the three connections on the side of the transformer.

High-power transmission lines distribute electricity across the country. Typical voltages are 138,000, 345,000 and 765,000.

A typical single-phase power line. The top wire on the insulator feeds all houses along the power line. The lower line is the neutral with a grounding wire tapped into it. The grounding wire continues down the pole to a ground rod.

The substation uses step-down transformers to lower the high-power transmission line voltage to a few thousand volts and distributes the lower voltage locally.

The Power Line

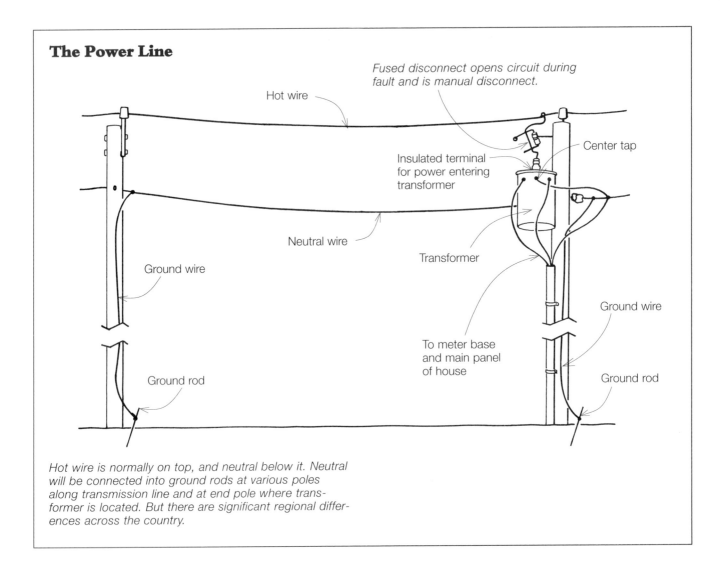

Fused disconnect opens circuit during fault and is manual disconnect.

Hot wire

Insulated terminal for power entering transformer

Center tap

Neutral wire

Transformer

Ground wire

To meter base and main panel of house

Ground wire

Ground rod

Ground rod

Hot wire is normally on top, and neutral below it. Neutral will be connected into ground rods at various poles along transmission line and at end pole where transformer is located. But there are significant regional differences across the country.

Let's take a closer look at the transformer (see the drawing on the facing page). It has an insulator on top to connect into the high-voltage (hot) utility line, which connects to the primary coil in the transformer. The bottom connection of the primary coil is connected to ground via the center-tap (CT) connection on the outside of the transformer. This makes a complete primary circuit that transfers its energy into the secondary coil, which has two hot ends and a CT in the middle. The CT connects to ground, along with the primary ground. Grounding the middle of the secondary coil splits the 240 volts in half so that we can have, in reference to ground or neutral, two 120-volt sources as well as one 240-volt source (120 volts from each insulated wire to the ground/neutral, and 240 volts from hot wire to hot wire). Connected to the three outside taps are three wires: two insulated hot wires and one bare ground/neutral. For an aerial feed, the two hot wires loop down and wrap around the ground/neutral for support to the house. This combo of three wires is called a service drop, or triplex (for more on wiring the service entrance, see Chapter 3). The service drop connects to the service entrance. For a buried feed, the two hot wires and the grounded neutral are run

Inside the Utility-Pole Transformer

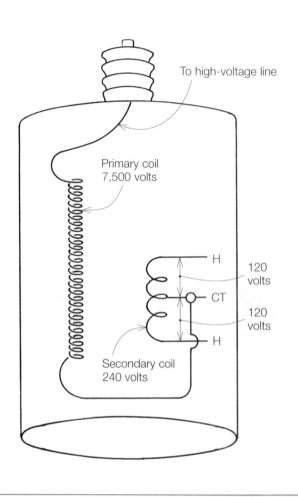

Utility-pole transformer has insulated terminal on top, which connects high-voltage line to primary coil (drawing facing page).

Primary coil is connected to ground/neutral via center tap, which then transfers its energy into secondary coil to step down voltage from 7,500 volts to 240 volts.

Grounding middle of secondary coil splits 240 volts in half, resulting in two 120-volt sources and one 240-volt source.

To high-voltage line

Primary coil
7,500 volts

H

120 volts

CT

120 volts

H

Secondary coil
240 volts

CT = Center tap
H = Hot

---

down the pole through conduit, into the ground, then they are connected to the service entrance of the residence.

ANATOMY OF A RESIDENTIAL ELECTRICAL SYSTEM

The house wiring begins at the service entrance. In this section I want to show you a typical house wiring system (see the drawing on p. 20). I'll go into more detail about all these circuits in later chapters. This example is just that, and it's important to remember that different houses require different

setups. I've already given you some background on where the power comes from. Now let's see how it all gets hooked up inside the house.

Let's go back to the service drop, or triplex (this example is an aerial feed). The utility wires splice onto the service-entrance wires next to the house at a drip loop. The utility's responsibility stops at the drip loop; from there, the owner's begins. Problems on the house side of the drip loop are paid by the homeowner. Problems on the utility side are paid by the utility. The masthead, which points down, and

The Main Wiring Runs in a Residence

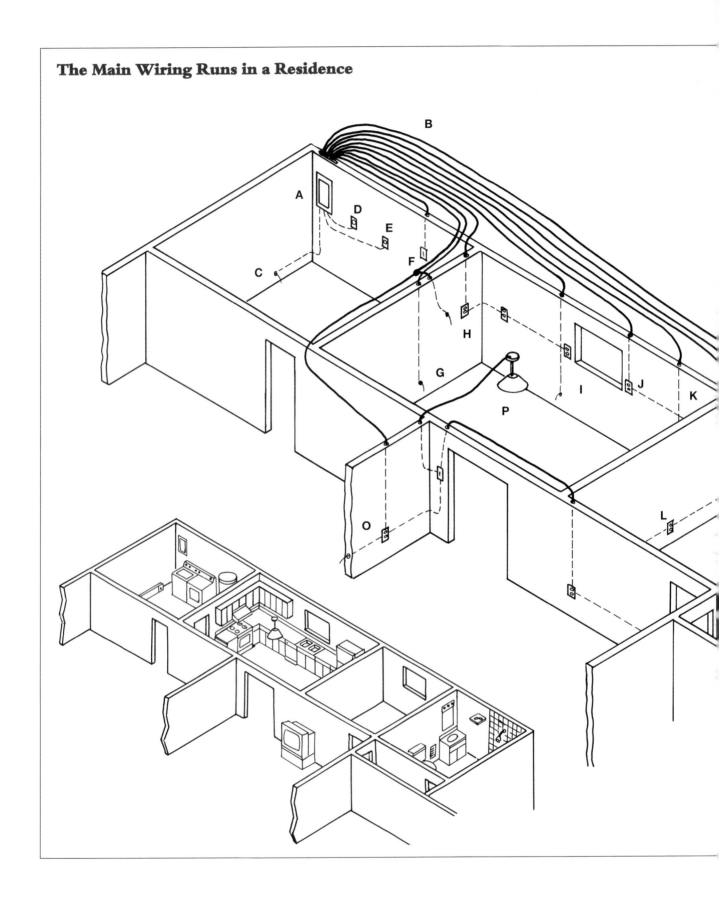

A. Main service panel
B. "Home-run" cables from main panel to loads (run through attic or ceiling or through basement)
C. Cable for baseboard heater (dedicated circuit)
D. Utility-room receptacle (dedicated circuit)
E. Dedicated circuit for dryer receptacle
F. Cutoff switch (optional when in sight of panel) for water heater (dedicated circuit)
G. Range cable (dedicated circuit)
H. Range fan feeds off living-room circuit
I. Dishwasher cable (dedicated circuit)
J. GFCI-protected receptacles on countertop
K. Refrigerator cable (dedicated circuit optional)
L. Cable for dining-room receptacles
M. Cable for in-wall heater (dedicated circuit)
N. Bath lighting circuit and GFCI-protected outlet
O. Cable for living-room receptacles and lighting
P. Kitchen overhead light powered off living-room circuit

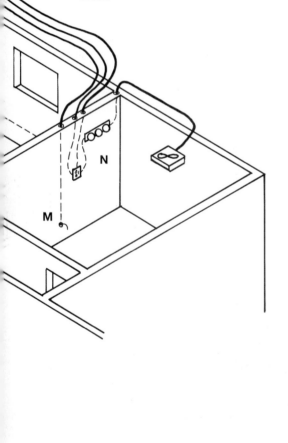

the drip loop, which circles up, keep water from following the wires, flowing into and down the mast and entering the meter base.

Once the service wires enter the masthead (called a weatherhead on the West Coast), they travel down the steel mast (called a riser on the West Coast) into the meter base (not shown in the drawing). From there, the SE cable goes directly into the main service panel. The main panel is where all wiring begins and travels to various receptacles, lights and appliances throughout the house. In addition, it's where the house electrical system obtains its earth ground via the ground rods.

To get the majority of cables out of the panel and into the house walls and attics, I cut a slot in the wall plate immediately above the panel, and I do the same in the bottom plate to get groups of cables into the basement or crawlspace. Now, here is where you have to be careful. When running cables through the house, keep them away from the edge of everything so that they don't get damaged by nails or screws. Even along a large wooden beam, the cables must be kept to the center.

The drawing at left shows some of the main wiring runs: the utility room, the kitchen, the living and dining rooms and the bathroom. These rooms are illustrated because each has its own special requirements. Cables going through walls are for several different functions, but, if needed, all of them can be routed together through the same holes in the studs. The cables in the utility room contain the 12-gauge utility-room cable (for the receptacle only), the four-conductor 10-gauge dryer cable, the 10-gauge cable for the water heater and a 12-gauge cable for the baseboard heater. Armored cable is used for the exposed run from the water-heater cutoff box to the heater. In this utility room, there is only one 120-volt receptacle, the one for the clothes washer: Most utility rooms will require more than one. Lighting for the utility room (not shown) is pulled off from the living-room circuit.

Deciphering Cable and Wire Codes

The only place you'll find more codes than on electrical conductors is in a spy novel. The gauge and manufacturer are not the only things stamped on the outside of a conductor. There is also a multidigit code that indicates what temperature range the conductor can physically be installed in. You don't want to put a conductor that's approved only for dry locations in a wet area (for example, underground), and you don't want to put a conductor that is approved only for low temperatures in an area with high temperatures, like an attic. Not only do you have to figure out the multidigit code, but you also must watch for preempt codes, which change or alter the previous code letters, depending on the installation.

The letter codes that appear after the gauge number and manufacturer's data refer to the insulation of the conductors. The letters tell you the type of insulation material, normally thermoplastic; the maximum temperature the insulation can be subjected to without damage, 60°C, 75°C or 90°C; and whether the conductor can be installed in wet or dry locations, or both.

Because the manufacturer is only trying to tell us three things, you would think it would be rather simple—but it's not. Besides the basic coding and the preempt coding, it also means something if something is missing (I'm not kidding).

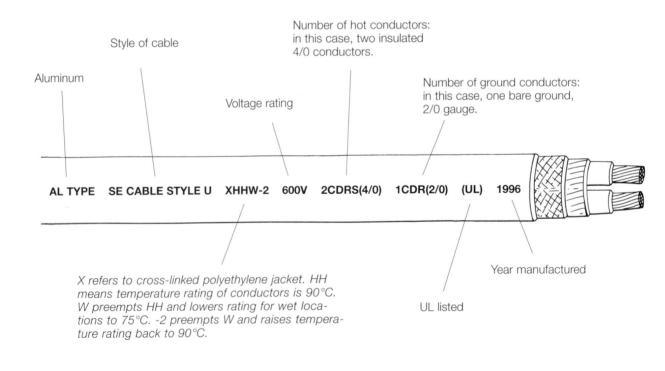

Aluminum — Style of cable — Voltage rating — Number of hot conductors: in this case, two insulated 4/0 conductors. — Number of ground conductors: in this case, one bare ground, 2/0 gauge.

AL TYPE   SE CABLE STYLE U   XHHW-2   600V   2CDRS(4/0)   1CDR(2/0)   (UL)   1996

X refers to cross-linked polyethylene jacket. HH means temperature rating of conductors is 90°C. W preempts HH and lowers rating for wet locations to 75°C. -2 preempts W and raises temperature rating back to 90°C.

UL listed

Year manufactured

SE Cable

Here's a sampling of the coding that appears on cables and other conductors commonly used in a residence:

H: If present, the maximum allowable temperature is 75°C. If not present, the maximum temperature allowed is 60°C. NM cable uses the letter B instead of H, but it works the same way: with B, 90°C, without B, 60°C.

HH: The maximum temperature is 90°C.

N: Nylon sheath around the thermoplastic insulation.

T: Thermoplastic insulation around the wire.

W: If present, the conductor can be used in wet *and* dry locations. If not present, it can only be used in dry locations (sometimes damp). If the letter W appears after HH—the W is a preempt coding—the

insulation's maximum temperature has been reduced to 75°C for wet locations (it is still 90°C for dry locations).

X: Cross-linked polyethylene insulation (normally used for SE cable).

-2: A preempt suffix that means the conductor can be used up to 90°C, wet or dry.

continued on p. 26

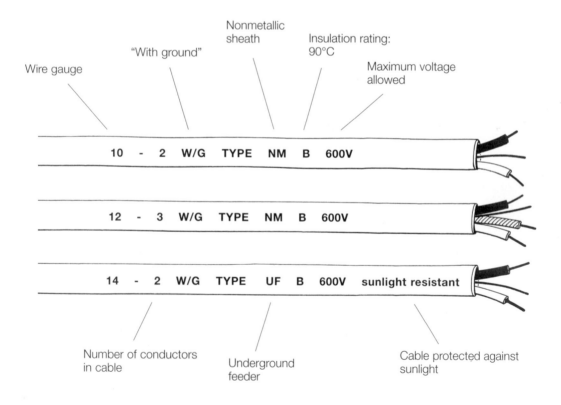

NM and UF Cables

Deciphering Cable and Wire Codes

continued from p. 25

Using the codes on p. 25, let's decipher THHN and THWN conductors. THHN is used in a home whenever conduit is required. It has thermoplastic insulated wire with an outer nylon sheath and can be used in dry areas where the temperature could reach 90°C. THWN, also common in homes and run in conduit, has thermoplastic insulated wires with a nylon sheath and a temperature rating of 75°C, in both wet and dry locations.

Though NM cable is fairly simple to decipher, SE cable is another story. For example, if SE cable has XHHW-2 stamped on it, the insulation is rated up to 90°C, but the W preempts the HH and indicates that the conductor can be used in wet locations down to 75°C. The -2 preempts the W and gives a 90°C rating to both wet and dry locations. (Confused yet?)

Although there are many codes and conductors, not all of them apply to typical residential applications. For a complete listing of codes, see section 310 of the NEC.

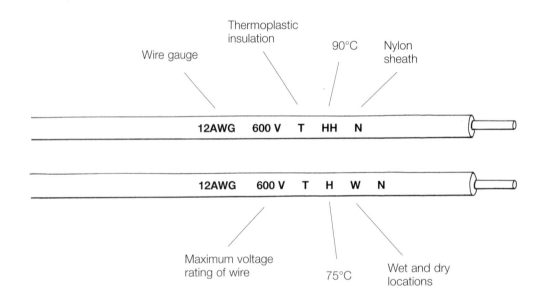

THHN and THWN Wires

The bathroom receptacle is powered off its own 12-gauge circuit, as required by the 1996 NEC. All outlets are protected by ground-fault circuit interrupters (GFCIs), and the in-wall heater has its own dedicated circuit.

All habitable rooms must have switched lighting at the point of entry. Receptacles and switches in the living and dining rooms are powered off individual circuits for each room. The cable enters through the ceiling and connects to a receptacle, which feeds other receptacles and switches in the room.

The kitchen circuits include two 12-gauge small-appliance circuits, the lighting circuit, the 6-gauge (four-conductor) stove cable, the 12-gauge dishwasher cable and a 15-amp refrigerator circuit as allowed by the '96 code. Receptacles along the countertop are GFCI-protected, as dictated by the '96 Code.

Be Careful When Running Groups of Cables

Running a large number of cables through joists or across the top of the attic beams can really get you into trouble. On one job that failed inspection, a man had all the cables bundled tightly from one side of the joists to the other. It looked beautiful: 15 to 20 cables all wrapped together going through the joists as neat as a pin. But such beauty has a major engineering drawback—the cables on the inside of the bundle will overheat because they cannot get any moving air. This will cause the insulation to break down, which can lead to the wires shorting together. Always spread the cables through several holes when running through joists or even studs, and when running the cables on top, spread them flat over a large area (for more on running wires, see Chapter 6). The attic cables are located over to the side where there is little headroom and less chance of someone tripping over them. Cables also must be stapled adjacent to the box they terminate into and along long runs.

Chapter 2

TOOLS OF THE TRADE

GENERAL TOOLS

AC-POWERED TOOLS

CORDLESS TOOLS

SPECIALIZED TOOLS

SAFETY EQUIPMENT

I can't remember the first tools I bought; it's been too long ago. But I do remember the first professional tools I bought and how comfortable they felt in my hands—well-balanced, well-designed and just begging to be used.

Buying quality tools is important, and so is maintaining them. You can judge how serious people are about their work by the quality and maintenance of their tools. Most professionals have broken more cheap tools than they care to think about and therefore have learned that cheap tools only get you into trouble. It's equally important to use the right tool when working with electricity. One of the first lessons an electrician learns is that the right tools can make a hard job easy and safe—and the wrong tools can make an easy job hard and dangerous. (To borrow from an old phrase: There are old electricians, and there are foolish electricians, but there are *no* old, foolish electricians.)

In this chapter I will tell you what I've learned about electrical tools from my many years of using (and abusing) them. And I will throw in a few tricks of the trade as well. I'll show you the tools an electrician uses every day, how to use them safely and how to maintain them to ensure a long life span—for both tool and user. And here's an important note: When on the job site, many tools look the same, and someone may claim yours. To help prevent this, put your name on all your tools with paint, stickers or by other means.

GENERAL TOOLS

General tools are the tools you'll be using most often. Though I can't cover every tool, I will discuss the ones that are most important to me, how to use them, and their advantages and disadvantages. Stay away from dime-store specials. Quality tools may be expensive, but they are worth it; if you don't buy quality electrical tools, you should buy a good, prepaid insurance policy.

I'm fastidious when caring for my tools. All tools require maintenance, even hand tools. Whenever a tool becomes worn or broken, fix it, or replace it immediately. Also, use the tool only for its intended purpose. Most hinged tools need a drop or two of lubricant about once a month. Some will need a lubricant spray to clean out the dust and debris. The lubricant should be nonsticky so that it won't attract dust, but it should leave a slick film to allow for smooth operation.

Insulated Handles

Most hand tools come with plastic-dipped handles, which have saved many electricians who have cut into a hot wire. But don't be misled: Even though plastic-dipped handles have saved many of us from unwanted electrical therapy, they are not designed to protect you against electrocution. For electrocution protection, look for special orange-colored insulated-handle tools with a double overlapping triangle symbol with "1000" stamped immediately to the right. This symbol means that the handle allegedly protects the user up to 1,000 volts. A hand tool designated as such has a high-dielectric white inner coating with a flame-retardant, impact-resistant, bright orange outer coating. An integral guard on pliers, cutters and strippers prevents your hand from sliding up the tool and contacting live wires. The tools available with insulated handles include screwdrivers, diagonal side-cutting pliers, long-nose cutters, pump pliers, cable cutters, strippers, nut drivers, wrenches and ratchets, and even a knife.

If you do accidentally cut into a hot wire using an insulated tool, the primary danger will be your reaction, not electrocution. Let's assume a worst-case condition: The wire is hot, and you did not know it. As the pliers cut into the wire, a direct short will occur from the hot wire through the cutter bar of the pliers and into one of the other wires.

There will be a loud arc and a flash. When this happens, freeze—do nothing—until the breaker has tripped. You will not be hurt. But you could panic and jerk away, then, if you're working off the ground, fall off a ladder or scaffolding. If you're using insulated tools and you cut into a hot wire, the only thing that will be ruined is the tool—if you don't panic. But even if you use insulated tools, always remember to verify that the power is off.

Insulated tools can be bought individually or in complete kits like this one. (Photo courtesy of Certified Insulated Products)

Side cutters, or lineman's pliers, are used to cut and pull electrical wire.

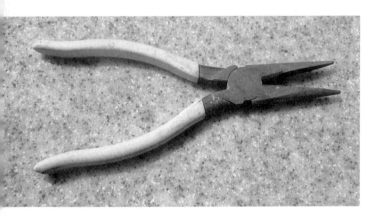

Long-nose pliers are indispensable for twisting wire and getting wire into narrow places. Their primary duty is twisting wire into a half-loop so that it can be attached to a switch or receptacle.

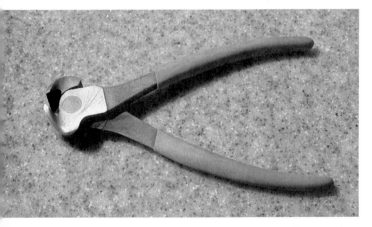

End-cutting pliers are normally used for pulling staples, but they can be used to cut wires installed straight in.

Tool Belt

The tools in each electrician's tool belt will vary slightly depending on their specialty. Each tool should have its own place in the belt so that it can be withdrawn without looking. The most-used hand tools in an electrician's arsenal are kept in the tool belt; I keep less-used tools in a bucket, which also serves as a seat.

Side Cutters

Side cutters (see the top photo at left), also called lineman's pliers, are to an electrician what a side arm is to a gunslinger of the Old West. Although used mostly for cutting and pulling wire, they can be used for other jobs too. The blunt, wide nose can twist wires together before a wire nut or any other mechanical splicing device is applied.

Good side cutters are among the most misused of the lineman's tools. Being heavy-duty, they are occasionally used for a hammer and other things I am ashamed to mention. The only way I stopped abusing my side cutters was by keeping a hammer and other miscellaneous tools by my side (normally in the bucket mentioned earlier).

Long-Nose Pliers

Next to my side cutters, long-nose pliers are the most used tool (see the middle photo at left). They are great for bending wires into a loop for insertion around receptacle and switch screws, and for pulling wires into narrow locations. Never use them to twist nuts off screws; the twisting action will spring the pliers' jaws off to one side and eventually ruin them. Try to find long-nose pliers with an integral wire cutter so that you won't have to change pliers just to cut the wire ends.

Diagonal Cutters

Sometimes wire needs to be cut extremely close to the surface of an object so that there is minimal wire protrusion. The thin head of a diagonal cutter allows it to fit into narrow places, where other pliers can't go. I often use mine when I have to cut one wire in a bird's nest of wires.

End-Cutting Pliers

Also called "nippers," I use end-cutting pliers primarily for pulling staples (see the bottom photo on the facing page). But they also make good straight-in close-cutting pliers. Sometimes a wire needs to be cut, but access is only straight in from the front. Side cutters and needle-nose pliers all cut from the side of the jaw, meaning that the entire head of the pliers must be at a right angle with the wire. End-cutting pliers cut dead-ahead.

Wire Strippers

There are many different types of wire strippers on the market (see the photo at right). If you don't have one, I strongly suggest getting one. Any wire stripper is good as long as it strips the insulation off the wire and doesn't nick or cut the metal of the wire. Try not to slice the insulation from solid wire with a utility knife. I know this is commonly done—I have done it, too. However, the knife blade notches and thus weakens the wire, which could break. (This normally isn't a problem with large-diameter wire.) If you must use a knife to trim insulation off a wire, cut as if you were slicing shavings off a piece of wood. Never circle a wire with the knife because it cuts a groove, which may make the wire snap when bent.

Of the various stripper designs, I use mostly the cut-and-pull type. My favorite used to be an automatic wire stripper, which cuts and strips off insulation in one operation without cutting into the wire. But automatic strippers don't seem to work well on the new NMB cable, which has an outer nylon sleeve on its individual insulated wires. Previous to that, the worse problem I had was that the tool wouldn't fit into a tight location. For such a situation, I used a cut-and-pull stripper. Now I use the cut-and-pull stripper all the time. Of the various strippers, always be sure the one you buy strips the most common wires: 16, 14, 12 and 10 gauge. The 16-gauge wires are used in light fixtures; the others are used in common house wiring. Simply place the stripper over the wire, squeeze and pull the insulation off the end of the wire.

Wire strippers trim insulation much faster and are easier on the wire than a utility knife is. The automatic wire stripper (right) strips up to 1 in. of insulation off the wire at a time. The cut-and-pull stripper (left) allows wire stripping in tight locations.

Screwdrivers

An electrician needs many different kinds of screwdrivers, and they all can't be kept in a tool belt. I keep one high-quality multipurpose screwdriver (a driver with interchangeable heads) in my pouch, and all the rest in the tool bucket.

There are many different types of screwdrivers: Flat head, Phillips head and offset are the most common used by electricians. Offsets are used in tight places, where a standard screwdriver is too long. I use mine to remove screws from an appliance, where there's only about 2 in. between the screw head and the wall. Don't use the wrong screwdriver for the wrong job. For instance, never use a straight blade in lieu of a Phillips head. And remember that even straight blades have different blade thicknesses. I keep an offset flat head and offset Phillips in my tool bucket for the times I have little clearance over the screw.

Screwdrivers, like side cutters, are often misused: I've seen them used as chisels, prybars and punches. To reduce the temptation of misusing your screwdrivers, carry the tools that you really need—the ones that the screwdriver is doubling for—in your tool bucket.

Always use screwdrivers that are in good condition and have insulated handles. Screwdrivers with worn-out blades and cracked handles have no place in an electrician's tool belt. Have at least one screwdriver with an insulated shaft because the day will come when a professional electrician (not an amateur) will need to tighten a hot screw. You can buy screwdrivers with a plastic sleeve over the shaft, or you can tape up the shaft on your own, as I do. But don't be prying around hot wires unless *both* the shaft and handle are insulated.

Utility Knife

I've yet to find a high-quality utility knife on the market, so I can't advise you on which one to buy. Most use the same cheap retraction mechanism that jams when it gets full of dust and dirt. But even with its problems, I prefer the utility knife over other types of knives because the blade is sharp and thin, and I can replace a dull blade fast.

You may prefer to use a standard knife instead. But on the whole, a general-purpose pocket knife makes a poor substitute for a utility knife. The point of a pocket knife is rarely razor sharp.

Stripping the sheath from a cable is an art within itself. You feel just like a surgeon; cutting the skin but nothing under the skin. Keep the cable flat, not twisted, and supported on a solid surface. Never, under any circumstances, support the cable with your knee, as I have done, while you're cutting the sheath. As soon as you think everything is fine, the blade will be buried deep in your leg (I have a scar to prove it). And don't try to cut into the sheath while holding it in mid-air. Once properly supported on a table or bench, use a *sharp* utility knife and gently slice down the center of the cable. The cable center contains only the bare ground wire, so there will be no harm if the blade occasionally goes in too far. Once cut, peel back the sheath and slice it away. You can try using cable sheath cutters (also called rippers)—I have a couple of my own—but I prefer the utility knife.

Electrical Tape

The electrician will need both black electrical tape and several different colored electrical tapes. White tape is used to identify any wire that is a neutral (grounded conductor) that doesn't already have a white coloring on its insulation. A gray color may be used for neutral marking, but I have never seen it done. Green tape can be used to identify grounding wires, the wires that ultimately connect to earth ground via a ground rod. Any color tape other than white (gray) or green, can identify different hot wires if you want to keep track of them. Colored tape should be used for color-coding only. Once you've used colored tape, you will realize that it is not the same quality as good black electrical tape. Black tape is used for general taping purposes—splicing, insulating a metal box, holding things together—and identifying a hot wire from the panel or utility.

I do not recommend using the cheap electrical tape found in grocery stores and discount houses. I prefer a high-quality tape, such as 3M Super 33+ (7 mil thick) or Super 88 (8.5 mil thick). Both are good tapes that work from 0°F to 260°F, resist ultraviolet rays and are UL listed (see p. 17 to find out what UL listed means).

For splicing in extremely hot applications, above the temperature range of standard tape, I recommend using 3M #69 Glass Cloth electrical tape. For covering splices with a solid rubber or silicone rubber tape, use 3M #130C or #70. All tapes are available through professional electrical suppliers.

Extension Cords

Extension cords are valuable additions to any person working in the trades. Cords come rounded or flat; I prefer the round because they are much easier to loop for storage.

Typical extension-cord gauges are, from thinnest to heaviest, 18, 16, 14, 12 and 10. A cord with a smaller gauge number, say 10, provides more power than a cord with a large number, say 14 (for more on wire gauge, see pp. 8-9). As a professional using heavy-duty tools, I never use any cord less than 12 gauge, but lighter-duty tools could use 14 gauge. The longer the cord needed, the heavier gauge you need. I have two 50-ft. 12-gauge cords, one 50-ft. 10-gauge cord and a 100-ft. 10-gauge cord. To keep the longest from getting tangled, I wind it around a 10-ft. 2x6 with Vs notched on both ends.

Make sure the cord you buy has enough outlets. You should either buy a cord with multiple outlets or buy a separate 1-ft. to 3-ft. cord with multiple outlets. Never use the multiple tap that plugs into an outlet—it can't take the current flow.

Some extension cords have ground-fault circuit interrupters (GFCIs). If yours doesn't, always be sure to plug the cord into a GFCI-protected outlet. Contractors are now required to use GFCI protection, whether the job is new construction or renovation.

Ladders

Ladders, like extension cords, are necessary for most trades. I have a 6-ft., an 8-ft. and a long extension ladder... all fiberglass. I chose fiberglass because it is strong, doesn't absorb water, doesn't warp, is corrosion-resistant and—this is very important—is nonconductive. Electricians should never use aluminum ladders unless they want to have a short life. Wood is fine but cannot take the abuse and weathering of fiberglass. My wood ladders lasted less than five years.

Ladders come in different grades and may be very confusing. All you have to remember is to get a type 1A fiberglass ladder of whatever length you want. (The 1A designation means that the ladder is extra-heavy duty.) Sometimes you can go by color to determine the ladder grade. Louisville ladders, for example, are color-coded: orange for class 1A and yellow for class 1.

There are all kinds of accessories available for ladders, and you will eventually buy a few of them as the need arises. For instance, because ladders will make scratch marks where they touch a house wall, inside or out, you may want to cap the end somehow. A low-cost method is to wrap a towel around the ladder ends that make contact with the siding or wall. Another option is to buy special end caps.

AC-POWERED TOOLS

Many types of AC-powered tools are required on the job site. Whether it is a drill or a saw, the extra torque provided by corded tools—as opposed to cordless, battery-operated tools— is a nice thing to have. Here, I will tell you my experiences with such tools, both bad and good.

Safety is paramount at home or on the job site, and tool maintenance is important to your safety. Follow the manufacturer's instructions for tool use and maintenance, and pay attention to the power cord on the tool. The cords on AC-powered tools get worn and damaged (I've lost count of the number of cords I've seen with damaged sheaths or bare wire showing). These should be fixed or replaced as soon as you notice any signs of wear. And although I shouldn't have to say this, never break the ground lug off a three-wire cord to fit a two-wire outlet—this is compromising safety for convenience. When the time comes to replace the cord's plug, install a high-quality one as opposed to a dime-store plug. Also, never use a power tool around water, or even excessive moisture, and always plug into an outlet protected with a GFCI (I'll talk more about GFCIs in Chapter 8). And never use a power tool that you don't feel comfortable with or intimidates you.

Drills

Roughing-in wiring in a residence requires heavy-duty drilling. Although most drilling can be done with a ⅜-in. drill and a sharp spade bit (for more on

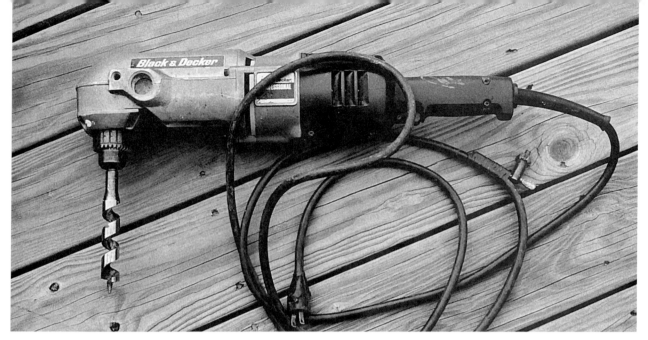

This ½-in. right angle drill stands up to heavy-duty, every-day use on the job.

bits, see pp. 35-37), a professional has a wide spectrum of drills and bits. I use two AC-powered drills: a ½-in. pistol grip and a heavy-duty ½-in. right-angle drill (see the photo above). A ⅜-in. drill just doesn't have enough power. Every-day, heavy-duty use would eventually ruin it.

Heavy-duty drilling rigs, however, are very dangerous and should be used only by professionals—and with extreme care. They have so much torque that, without a clutch, they can rip off an arm or break a wrist. I currently use a Black and Decker TimberWolf, which has a clutch, on low speed, that will engage when the bit sticks. It has two speeds: 300 rpm for drilling large-diameter holes; and 1,200 rpm for speed drilling. I use the low speed whenever I am using a self-feed or a hole-cutting bit and sometimes when I'm using an auger bit. This drill is what you want all drills to be: heavy duty and of good design. Right-angle drills also come in medium duty, which also can injure you. I do not recommend them for the professional because they will invariably get used in heavy-duty situations. My pistol-grip drill is a Makita ½-in., but don't be deceived by its look: Being ½ in. and without a clutch, it can break your wrist in a heartbeat just as a right-angle drill could.

To operate heavy-duty drills safely, follow these precautions:

1. Maintain a good, tight grip on the tool.

2. Always assume the drill may twist out and fly around. So keep your head away from anyplace where it could be smashed by the drill if it gets away, and don't hold your arms in an awkward position. Also, do not stand in a precarious position to drill; anchor yourself in a good, solid position.

3. Wear heavy leather gloves and safety glasses.

4. Feel what is happening through the vibration of the drill. Many times you can pull back as you feel the drill starting to work harder or get jammed up on something or even hit a nail.

5. Never drill with a damaged or dull bit.

6. If you feel the bit jam, let go of the trigger immediately. Try not to let go of the drill. If the drill twists out of your hand, jump out of the way immediately, which leads to the next rule.

7. Always have a clear area behind you.

Drill Bits You'll Need

For cutting through wood, I use several different types of bits. Which bit you use will depend on the situation, whether you'll need to cut through a double top plate, through a corner with nails or simply through studs. It is important that you know what each bit can and cannot do so that you can choose the right bit for the right job.

Spade bits
There are several different spade designs on the market. I've found that spade bits with two protruding end cutters cut faster than the standard flat-bladed variety (see the photo at right). Spade bits come in short (4-in., but I've only seen these in DeWalt's line), standard (6-in.) and long (16-in.) lengths. One problem with spade bits is that they tend to break at the shank if you don't keep the drill centered in the cutting hole—DeWalt solved that problem by making its shanks extra-thick. And DeWalt makes a ¼-in. hex shank attachment for the hex drivers, which allows the blade to be installed without opening the chuck. This is especially nice for the keyless chuck, battery-operated drills because you no longer have to waste battery power (stalling the drill out) to change the bits. Spade bits normally come in ¼-in. to 1½-in. cutting diameters.

Auger bits
Auger bits are long-bladed spiral bits used to drill long distances through wood or through several studs at once (see the photo below). One problem with long drill bits, and especially auger bits, is that they tend to bind and get stuck in the hole. When this happens, remove the jammed bit. First unplug the drill and remove the drill from the bit. Then with a small pipe wrench, turn the drill-bit shaft counterclockwise to unscrew the bit. To prevent this, coat the drill bit's flat sides with wax, soap or wire-pulling compound. I find this especially useful when I'm drilling at an angle or very deep into the wood. The slick coating keeps the bit from binding on the sides of the hole. Another problem with auger bits is cutting into nails: You'll normally feel a repeating thunk

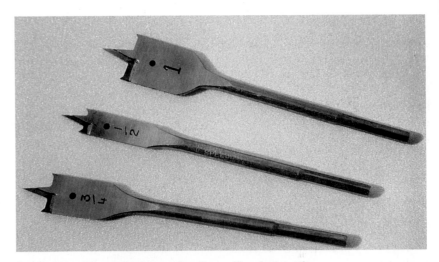
Spade bits with protruding end cutters allow fast cutting.

Auger bits, both short and long, are used for fast and deep hole cutting.

Drill Bits You'll Need

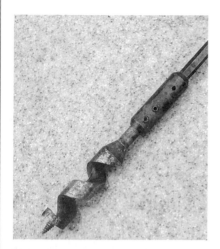

Extension arms can make bits drill deeper into the wood. This 12-in. extension attaches to the auger bit via two set screws.

as the edge of the cutter keeps cutting into the steel. This won't be a problem if you're using a bit that can cut through steel, like the Greenlee NailEater. If you hit a nail with this bit, the auger will cut through it, as opposed to jamming the bit and possibly hurting the person holding the drill. (Before drilling, try to remove any nails that will be in the way or find another route.)

Auger bits are available in different diameters (¼ in. to 1½ in.) and lengths (6 in., 7⅝ in. or 7¾ in. and 18 in.). I recommend both a long bit (18 in.) and a short bit (6 in.), as well as an extension (see the photo above). I use 1-in., ¾-in. and ½-in. diameter bits.

Be extremely careful when using an auger bit in a pistol-grip drill (I don't recommend it): The torque could make the drill twist with the bit, possibly breaking your wrist. To counter this, most manufacturers make a side handle that fits into the drill for extra holding leverage.

Self-feed wood borers
The large boring bits that cut away all the wood within the perimeter of the cutting circle are called self-feed wood borers (see the photo below). These bits come in either one solid piece or with blades—I prefer the blades because they cut very fast and accurately. However, if either cuts into a nail, look out! The bit simply stops turning. And when this happens, the drill body itself will start to turn, hitting the hapless worker in the head or breaking his/her arm. This type of bit should only be used by professionals with a right-angle drill—preferably in a drill with a clutch.

Hole saws
The bits that cut only the outside perimeter of the cutting hole are called hole cutters, or hole saws. I started using hole saws in lieu of self-feed bits because I had so many close calls with accidents. I now use only carbide-tipped hole saws for opening large holes (they are also available in bimetal). I prefer the carbide tooth because it can cut through a nail as if the nail wasn't even there. A nail will ruin a bimetal cutter.

Self-feed wood borers cut large-diameter holes fast. The bit on the left has replaceable cutting blades. Such bits should only be used in a heavy-duty, right-angle drill.

Hole saws need an arbor, which is separate from the cutter head (see the photo below). The arbor fastens to the cutter head, then fits into the drill chuck—it transfers the drill's rotating power to the cutter head. One arbor can be used with a number of hole saws, which keeps the cost of the cutter heads to a minimum. Carbide-tipped hole saws are available in sizes from ¾ in. to 6 in.

An option to the two-part cutter-head and arbor system is the one-piece arbored hole saw. These hole saws may be more expensive because you have to buy the arbor each time you buy the hole saw. I don't use them much because they are normally only bimetal tipped.

Twist drills are commonly used for both wood and steel drilling. The two on the right are cobalt-coated to keep the sharp edge longer.

Hole saws are safer to use than self-feed bits. They come either with or without an arbor.

Either system, whether integral arbor or separate, will require a pilot bit, which centers the hole saw in the hole to be cut. Pilot bits have a habit of breaking, so have several spares on hand. Lennox recommends using a carbide-tipped pilot bit—I do not. A standard twist-drill bit will cut into the wood faster and will be cheaper to replace if the bit breaks.

Twist drills
Standard steel-drilling bits (called twist drills) are needed for drilling wood and steel. A common application is for drilling pilot holes through a floor and into the crawlspace to act as a position locator. For this job, a ⅛-in. to ¼-in. bit, around 12 in. long, works fine. I've found I don't need the

complete spectrum of bits down to 1/32 in. that get sold in complete kits. I normally use 1/16 in., ⅛ in., ¼ in., ⅜ in., ½ in. and ¾ in. Needing fewer bits, I treat myself to buying bits with the gold-colored cobalt coating so that they will stay sharp for a longer period of time (see the photo above). Never drill pieces of metal, or anything for that matter, while the piece is not supported firmly (clamped, for instance) or stationary. One time I thought I could drill through a small piece of metal while I was holding it. The bit caught the metal, the metal turned in a circle with the bit, and it gouged out a very large and deep hole under my thumbnail. Please learn from my mistakes; don't repeat them.

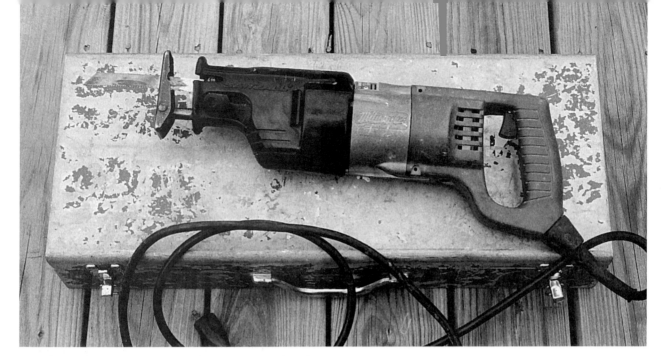

A reciprocating saw will cut anything, from wood, nails, steel and cast-iron pipe, and rebar.

Saws

Two saws I use on the job for rough-cutting are the reciprocating saw and the jigsaw. My primary tool for cutting anything is the reciprocating saw (see the photo above), commonly called a Sawzall, named after the original design by Milwaukee. Currently, I'm using the Milwaukee low-vibration design Super Sawzall (6527). I've been using it for several years, under hard work and neglect, with no problems. I cut wood, nails, steel and cast-iron pipe, rebar— anything I want. It is with the reciprocating saw that you'll make most of your accidental cuts through hot wires. This is why it is very important to have a double-insulated tool, or as in the Milwaukee, a grounded tool. In addition, the Milwaukee has a rubber boot over the front half that you hold when pushing the blade into the wood. I've cut many hot wires without even a minor shock. I've lost only the blade—and the wire.

Besides the standard blades for reciprocating saws, extra-thick and bimetal types are available, all in varying lengths. The reciprocating saw makes only rough cuts because the blade tends to move around in the cut. Some blades move more than others;

extra-thick blades keep straight in a cut better than any other. And some blades last longer than others. I normally use the bimetal type because these blades have some bend in them; they give a bit without immediately breaking. This type of blade is only slightly higher in price, but its longer life more than repays the investment.

A good jigsaw is another important tool in the electrician's arsenal. I don't like any jigsaw that uses the standard hole-in-the-blade-type blade design. The blades seem to break right where the hole is. I prefer a jigsaw that uses the bayonet design—these break, too, but not as easily. You'll need a variety of blades, and you'll break fewer of them, and be safer to boot, if you let the jigsaw cut at its own speed, with only slight pressure from you.

Never cut with any saw, especially ones with long blades, unless you know what is behind the wall. It could be plumbing lines or electric cable—even the main service-entrance cable. If you cut into the latter, you and the saw will be in big trouble because there may not be any overcurrent device (breaker) in the system, except what the utility has on the pole.

CORDLESS TOOLS

Battery-operated tools are flooding the marketplace. My first experience with them was distasteful: The first cordless tools on the market were ill-designed and useful only for very light-duty applications. This distaste lasted for many years. But today the market is full of high-quality cordless tools. Now I'm not without my battery-operated drill, saw and screwdriver. Cordless tools offer a few advantages specific to each tool. The most obvious, of course, is that the tools don't require an extension cord, so they are very portable and won't electrocute you.

All cordless tools need battery chargers (see the photo at right). In the past, the battery needed about an hour to charge—today a charger requires only about 15 minutes to do its job. For continuous work, you'll need several batteries on hand. I'd stay away from the charger design that requires you to push a button to start the charging cycle. You may come back later to find the battery isn't charged because you forgot to push the button, or you pushed it, and it didn't work. A good charger will indicate when it's charging, already charged and when it's done. It will also analyze the battery to see if it has any problems. For example, it will tell you if the battery is weak and cannot supply full power.

Drills

Cordless drills, when used with a sharp bit, can drill most of the holes in the house and won't break your arm doing it, as a heavy-duty AC model could. And they can't electrocute you, even if they cut into a hot wire, because the case is made of plastic. I do at least 60% of my drilling with cordless drills. They offer me speed, versatility and safety: I can use a cordless drill on a ladder, around water, and in crawlspaces and other awkward places where I would normally have to worry about pulling, dragging or lifting an extension cord.

Cordless drills are available in several different voltages—6.9 volts to 14.4 volts and higher. The 6.9- and 9.6-volt units are the current favorite for the

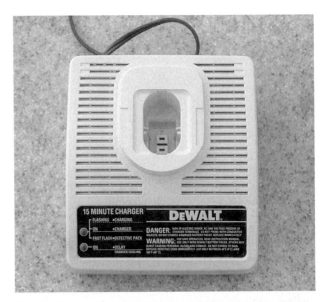

This DeWalt battery charger does its job in approximately 15 minutes. Simply insert the battery. The charger beeps when finished and will analyze the battery for problems and will tell you if it's already charged.

average do-it-yourselfer because they are lightweight and have a decent amount of power. The more serious do-it-yourselfer, or the professional, may need a higher-voltage, larger chuck unit.

Today's cordless drills are available with ⅜-in. and ½-in. chucks. I use the ½ in. because it can accept larger-diameter bits. Most cordless drills now come with keyless chucks. At first I couldn't wait to get them, and now I wish I hadn't. Keyless chucks are great for fast bit-changing, but be aware of their minor problems. Without using a chuck key, you can get the bit only "so tight"—and if you're drilling deep, the bit may stick in the wood and start spinning in the chuck. In addition, to tighten bit in the drill, you have to place your hand on the drill chuck head and pull the trigger. The chuck tightens until the motor stalls. Once the motor stalls, you have to assume the bit is tight in the chuck. This process wastes valuable battery power.

A cordless drill is great for working in a crawlspace, on a ladder and for most other light- to medium-duty drilling.

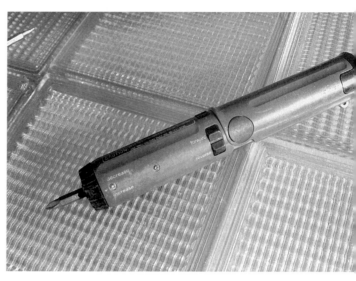

A high-quality battery-operated screwdriver is a must for removing switches and receptacles. The author's Milwaukee unit runs smooth and quiet with lots of power.

A battery-operated circular saw allows you to work in confined and damp areas. The thin, 5⅜-in. blade makes fast cuts without draining the battery.

My favorite cordless drill, and the one I currently use, is the DeWalt, which is a very heavy-duty model with a ½-in. chuck, electric brake, 15-minute charge and a 14.4-volt battery (see the top left photo on the facing page). In addition, there's an 800 number right on the drill if you have any complaints.

There are several kits available for cordless drills that make my work easier. One I'm especially in love with is a hex-head drill-bit set. The bits fit into a magnetic extension holder, which fits into the chuck. The extension allows me to go from drilling holes to driving screws without opening and closing the chuck, which saves time, trouble and battery power. Longer extensions allow me to work in tight locations where the drill head won't fit.

Circular Saw

Another battery-operated tool I use often is the DeWalt cordless circular saw (see the bottom photo on the facing page). This is not a toy—it cuts with the best of them. It works like a champ to open up areas for working, or for fast trims and cuts. It's nice just to pick the saw up and cut rather than find a grounded outlet and drag out the extension cord. If you're working in areas where no electricity is available, this is a good saw to have around.

Screwdriver

I love the Milwaukee 6538 cordless screwdriver (see the top right photo on the facing page). In my opinion it's the best on the market. It's heavy-duty, unlike the other play-toys sold at hardware and cut-rate stores. I use it primarily for getting into multigang switch and receptacle boxes. A four-gang switch box will have 16 screws to pull before entry into the box. Thank goodness for the Milwaukee.

SPECIALIZED TOOLS

The tools in this section are very specialized. I don't recommend that everyone doing minor wiring jobs buy them. Instead, these tools are for professional electricians and serious do-it-yourselfers (those folks who plan on doing some big-time wiring work around the home).

Hole Punchers and Stepped Drill Bits

All electricians know that the prepunched holes in meter bases and control panels are rarely where you want them. A lot of labor and material can be saved if you place the conduit and the cables where you need them, as opposed to where some desk jockey decided you needed them. There are many different ways to punch holes in metal boxes. The way I have settled into is with a 400:1 gear-reduction unit made by Cobra that uses Greenlee Slug-Buster hole punchers. First drill a pilot hole through the steel. Then the gear-reduction shaft fits through the hole with the Slug-Buster on each side (see the photos on p. 42). Attach a drill to the unit's shaft and pull the trigger. You can even use a cordless drill for this operation. The rotating shaft pulls the Slug-Buster together and punches the hole faster than I can explain it. I use a standard hole-punching kit that goes from ½-in. to 1¼-in. I also purchased the 1½-in. and 2-in. Slug-Busters. As opposed to punching holes, special hole-cutter-type drill bits are available to cut straight through the metal.

For drilling small-diameter holes quickly, such as the pilot hole for the hole puncher or a knockout, nothing beats a stepped drill bit. This is a special metal-cutting bit that cuts various diameter holes depending on how far you push the bit into the metal (the farther you drill, the bigger the hole). Stepped drill bits are available in two designs: a blunt-end variety used to increase the diameter of existing holes, and one that is tapered to a point to open new holes. I use either one to increase the diameter of the knockout holes (up to 1⅜ in.) in metal boxes. I also keep carbide-tipped hole cutters around for very thick steel.

Conduit Benders

Professionals will bend conduit, and do-it-yourselfers will use premade angle fittings. I keep ½-in. and ¾-in. benders on the truck for just those times when I'm allowed to have fun... I mean bend conduit. However, mastering the art of bending conduit takes a tremendous amount of time, patience and money (for wasted conduit to practice with). If you want to install the conduit fast, forego the conduit benders and use preformed angles.

Fish Tapes

Pulling and fishing wire can be a major problem if the wrong equipment is used. The purpose of a fish tape is to "fish" wire through walls. Here's how: Put the thin metal tape into a hole in a wall, maneuver it through a wall and bring it out a different hole. Then attach a wire to the end of the fish tape and pull the wire into the wall as you pull the fish tape out.

However, a standard metal-coiled fish tape is, in my opinion, the wrong equipment to use. It is conductive, has an end that catches on everything in the wall, so you can't get it back out, and is hard to get through a wall or ceiling cavity because it always tries to go in a circle. To counter this, I cut several feet of metal fish tape off my reel, bent it straight (I store it straight as well), cut the looped end off it and taped it with electrical tape. It now slides through the wall pretty easily. I don't recommend any fish tape made of metal, and the nylon tapes I've seen stay circular just like the metal ones and catch stuff in the wall cavity. However, I hear the fiberglass units work well.

What makes a good fish tape? It should be a nonconductor, flexible but just rigid enough to stay straight in the wall. A 20-ft. straight length of ½-in. or ¾-in polybutylene plumbing pipe makes a perfect fish tape. It is rigid enough to go long distances within walls and ceilings and stays flexible enough to bend into the cutout hole and pull through. Once through, you can push the wire up into the pipe and tape it off. As you pull the pipe back, so comes the wire. This is the method I use most often. I normally use ½-in. in the walls, and ¾ in. for the extra-long

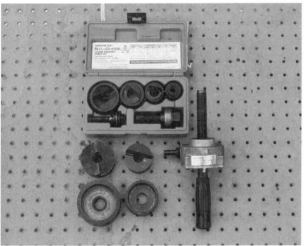

To punch holes in metal boxes quickly, the author uses Slug-Busters (bottom). Drill a pilot hole through the steel and put the gear-reduction unit through the hole with the Slug-Buster on each side (top). Then attach a drill to the unit's shaft and pull the trigger.

runs in the ceiling. To prevent the end of the pipe from catching, I tape or insert a marble to the end.

Rotary Hammer

Used to break up and drill concrete and punch through cinder block and brick walls, the rotary hammer (also called a hammer drill) is an indispensable tool for the professional. Whatever brand you get, make sure you get one with a clutch that engages when the bit sticks, because this type of drill has tremendous torque. I also use the drill for installing ground rods. Makita, Bosh and several other manufacturers make a special attachment for their hammer drills that will drive a ground rod into the soil in just a few seconds. It sure beats using the old sledge hammer.

VOMs and Multimeters

Volt-ohm-meters are the most valuable tools in any electrician's tool chest. I still call them VOMs because that's what we used to call them in the early days. Technically, in this day and age, they're called multimeters. Having been through many of these devices, from the old analog Simpson 260s to today's digitals, I have to say my digital Fluke model 25 and model 30 are the best I've ever used (see the photo at right). They do everything with extreme accuracy, and they are almost indestructible (I've dropped mine on concrete, and they still work). Some digitals make you wait while they count up to get the voltage. However, both of these models give an instant voltage reading. They also work in cold weather (some don't).

The model 25 is autoranging, meaning I don't have to know the voltage before I measure it and switch it to the right scale. I simply put the probes on the testing points. It has a continuity tester built in to test light bulbs and water-heater elements, capacitance tester to test motor starting capacitors, and a millivolt meter, which is useful when testing thermocouples on gas water heaters.

The model 30 clamp-on meter indicates how much current the branch circuit is pulling without having to open the circuit. I can check submersible pumps, water heaters and electric baseboard heaters to verify

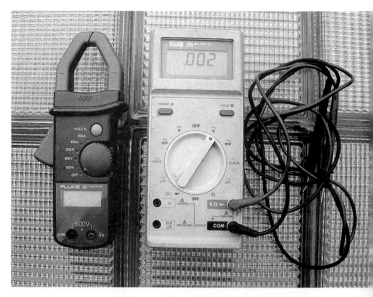

An electrician is only as good as his or her volt meter. Without it, he or she can do very little. These two meters, made by Fluke, are the author's favorites.

how much current they're pulling without getting near a bare wire. The model 30 is also a multimeter, measuring voltage, resistance and continuity (measuring continuity is a way of verifying that two points are electrically connected). However, it is not autoranging, meaning that you'll have to know the approximate voltage of the circuit, and then switch it to the proper scale. I prefer having two multimeters on hand—one as a backup.

There are many accessories available for VOMs: extra-long test leads, alligator clips that allow one or both hands to be free, and line splitters. Line splitters allow current to be checked in any corded appliance (see the top photo on p. 44).

Plug-in Analyzers an GFCI Testers

I use plug-in circuit analyzers and GFCI testers all the time to tell if the receptacle is wired properly and to see if the GFCIs still work—and you'd be surprised to know how many are not. However, shop around before you buy. Some test for five problems, others for six, and none can test for everything.

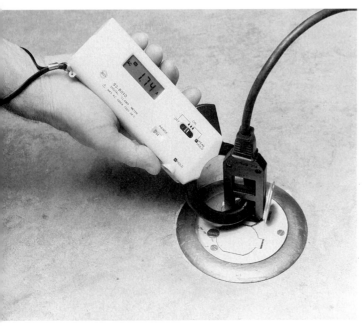

A line splitter allows you to check the current draw of any appliance without cutting into the wiring. It plugs into the outlet, and the load plugs into the splitter.

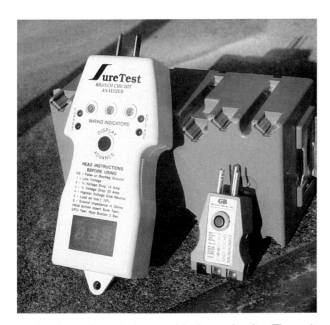

A plug-in analyzer helps troubleshoot circuits. The unit on the right only indicates whether a receptacle is wired properly. The circuit analyzer by SureTest (left) tests the entire branch circuit and looks for wiring errors.

A typical, low-cost plug-in circuit analyzer will give you the following readings:

Open ground Either someone didn't hook it up or the ground wire is broken.

Open neutral If the neutral wire was open, normally the outlet would be electrically dead. There could be several reasons for this situation, but it most often occurs when a neutral wire has pulled out of a push-in termination—either at the receptacle being tested, or the one preceding it.

Open hot Often occurs when a hot wire has pulled out of a push-in terminal—either at the receptacle being tested, or the one preceding it.

Hot/ground reversed Very rare indication, one that I've never seen.

Neutral/hot reversed The two insulated wires are switched on the receptacle.

Correct wiring The circuit is okay.

In the past, a GFCI tester was simply a plug-in tester with a push button on it to simulate a ground fault. No more. There is a new kid on the block. State-of-the-art testers can now give us a whole world of circuit analysis from a little plug-in detector. The one I use is the SureTest Digital Wiring Analyzer (see the photo at left). This unit is too elaborate and expensive for the amateur but should be in the arsenal of every electrician and official who does testing. Beside standard wire checks and GFCI testing, the SureTest looks for a boot-leg ground, where the neutral is wired over to the ground; low voltage and high voltage in a circuit; voltage drops in both 15- and 20-amp circuits; voltage drop from ground to neutral; and ground resistance and continuity from the testing point to the bond in the main panel.

SAFETY EQUIPMENT

The most dangerous problem on the job site—besides electrocution—is debris going into your eyes, which makes approved safety glasses an electrician's best friend. They must be comfortable, or they will not be worn. Be sure to get some that have adjustable arms. Antifog and antiscratch coatings are nice, but they fog and scratch anyway. I hang my glasses from the rear-view mirror of my truck, which reminds me to put them on when I leave the truck.

After your eyes, your toes are the most in danger. Always wear steel-toed shoes. Falling drywall, plywood and heavy control-panel lids are always threatening to amputate your toes or injure your feet. If you're around a job site long enough, something will happen. If hard-soled steel-toed shoes are not to your liking, steel-toed running-type shoes are now available.

Rubber gloves, leather gloves, a rubber mat and a hard hat are items that should be kept with you. Rubber gloves can save your life. Once I was wiring merrily away in a panel—no where near a bare hot wire—and zap! What happened? I brushed against a hot wire where a previous electrician had gouged off a section of insulation. From that time on, I started wearing rubber gloves in hot electrical panels or even when checking out a hot receptacle or switch. The rubber mat will insulate you from ground.

Chapter 3
THE SERVICE ENTRANCE

LOCATING
THE METER BASE

AERIAL SERVICE
ENTRANCE

BURIED SERVICE
ENTRANCE

CONNECTING METER
BASE TO MAIN PANEL

The service entrance is that part of the electrical system that starts at the utility transformer and terminates at the main service panel. It includes the wire from the transformer to the residence, the meter base and the service panel. A residence has either an aerial or buried service entrance. In an aerial service, the wiring that swings from the utility pole to the house is called a service drop, or triplex, and is provided and installed by the utility company (see the top drawing on the facing page). The utility's responsibility ends at the splice at the drip loop. From the splice, it is the electrician that provides and installs service-entrance (SE) cable to the meter base, and from there to the main service panel. In a buried service entrance, however, the utility is responsible for wiring from the transformer all the way to the meter base (see the bottom drawing on the facing page). The electrician installs the meter base and the SE cable and conduit from the base to the main service panel.

The first things to consider in designing a service entrance are the requirements of the utility and the local inspector, which may change from utility to utility and from locality to locality. Although the NEC is a national code, local requirements may differ and will supersede the NEC, so pay particular attention to odd-ball local codes: I've read of one town that requires the meter to be located within 2 ft. of a sidewalk. This may have you installing a sidewalk to the rear of your house if you don't want the utility meter next to your front door. And some municipalities require the SE cable to enter the building at a certain distance above grade if the

Aerial Service Entrance

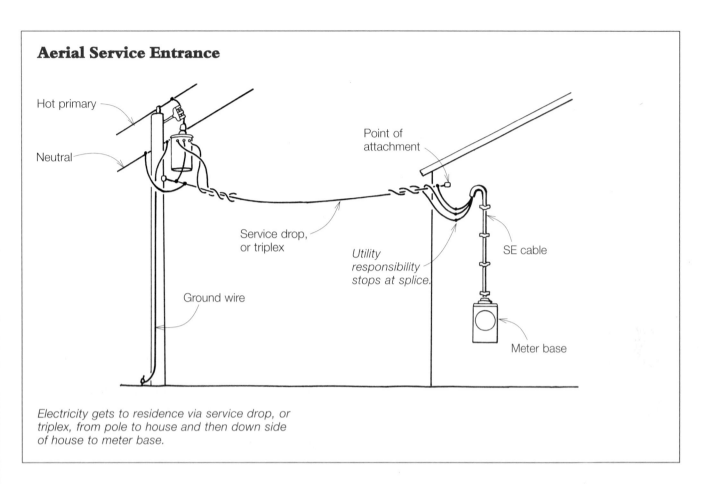

Hot primary

Neutral

Point of attachment

Service drop, or triplex

Utility responsibility stops at splice.

SE cable

Ground wire

Meter base

Electricity gets to residence via service drop, or triplex, from pole to house and then down side of house to meter base.

Buried Service Entrance

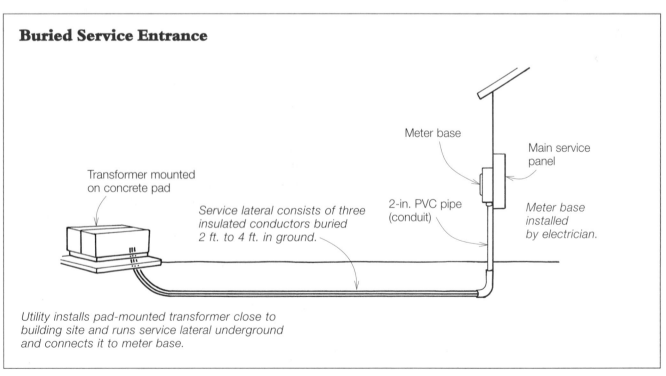

Meter base

Main service panel

Transformer mounted on concrete pad

Service lateral consists of three insulated conductors buried 2 ft. to 4 ft. in ground.

2-in. PVC pipe (conduit)

Meter base installed by electrician.

Utility installs pad-mounted transformer close to building site and runs service lateral underground and connects it to meter base.

main panel is to be located in the basement. The bottom line is, check with your local inspector at the design stage.

Another consideration is the size of the residence's service. It's important that you don't undersize the service entrance. You should leave room for expansion. Also, SE conductors must match the ampacity of the house electrical system. For example, if a 200-amp main panel goes in the residence, the SE conductor must be able to withstand 200 amps of current (for more on amperage ratings of conductors, see p. 9). If more than 200 amps is required, a large 400-amp meter base will be needed, which has extra or extra-large lugs that allow you to run parallel-service wires into the service panels. For example, a 400-amp service would consist of two 200-amp panels (a 300-amp main panel would consist of one 200-amp panel and one 100-amp panel) within the residence, with each panel wired with an individual set of conductors from the 400-amp meter base (for more on installing a 400-amp service, see p. 56). Setting up the service entrance this way makes wiring easier because you can use smaller-gauge wires (wires larger than 4/0 are awfully hard to handle in a residential situation because they're big and don't bend easily).

Many of the initial questions in new residential construction concern the service entrance, which is a major factor in the design and cost of a home. All through the design and construction of a house, it's important that there be coordination and cooperation between the utility, the local inspector, the building and electrical contractors and the homeowner to avoid problems and errors. This chapter will help you make intelligent decisions regarding the design of the service entrance.

LOCATING THE METER BASE

Before the utility runs the service drop to the house, you must install the meter base. If the meter base has to be installed before the finished siding is up, I attach the meter base to the rough wall and allow the finished siding to be cut around it. Another

option is to shim out to the exact distance of the proposed siding and install the siding behind it later. If brick siding is to be installed on the house, the meter base will have to wait for the brick to be installed first. You will find it difficult to drill into brick. Instead, try to fasten the box into the concrete between individual bricks. Any type of concrete fastener will work.

Regardless of whether you choose an aerial or buried service entrance, always try to locate the utility meter base and the main service panel back to back. This design is always less expensive and will be less of a problem when it comes time to run cable to the service panel (see p. 60). If the panel is installed several feet away from the meter base, a cutoff panel (which is expensive and will require additional work to install) will normally have to be installed next to the meter base, and there will be the cost of extra cable. The cutoff panel is simply a main disconnect switch, or circuit breaker, that provides fault protection for the SE cable on its way to the main service panel. Without a cutoff panel, if the cable is damaged—say, by a carpenter driving a nail into it—before it reaches the breaker, there is no way to shut off the power, other than the utility cutoff at the power pole. By the time the utility cutoff removes the power, all the service cable will probably be damaged beyond repair, and a fire may result. When such a fault occurs, the wires are humming loudly, the utility meter power-usage disk is whirling so fast you can't see it, and the utility transformer is buzzing. The need for the extra panel will have to be verified by the local inspector. If the cutoff panel is needed, code requires the main service panel to be wired as a subpanel. That is, all the neutrals must be isolated from the grounds (I'll talk more about subpanels on pp. 70-73).

Mounting the Meter Base

The meter base is either mounted at the exact height required by the utility (around 5 ft.) from final grade, or in absence of a spec, at head height for easy reading. Before you begin work, double check to be sure you have the right meter base—buried for buried, aerial for aerial, and the right amperage for the job being done.

In the meter base, you'll see four ¼-in. knockouts for four screws. Use a punch to remove the knockouts and be sure to remove all of the metal (check behind the box because it may bend around back).

Remove a knockout from the back of the meter base for the conduit that will be the pass-through for the cables running to the main service panel. (For the purposes of this discussion, assume the main panel is located back-to-back with the meter base.) Hold the meter base up against the building with one hand and trace the knockout hole onto the siding with the other. If the proposed hole will cut into a stud, move the base left or right to go around it. Cut the hole 1 in. larger than the circular mark. To be sure the connection between the base and the hole remains watertight, place plumber's putty or caulk on the building around the hole.

Now the meter base is ready to be attached to the building. (If an NM connector is going to be used in lieu of conduit, you may want to install it before you screw the meter base on; conduit can be installed after that.) Place a magnetic torpedo level on the meter base so that level can be observed at all times. Making sure the base remains level, align the meter-base hole, and screw the meter base to the building with four ⁵⁄₁₆-in. hex heads (don't use ¼-in. hex heads; they break too easily). Never use nails to attach a meter base; they can work loose over time. Once installed, I normally put a bead of silicon rubber along the top and sides of the base to keep water from running behind it. With the meter base installed, you're ready to run SE cable to the service drop (aerial service). For buried service, there is no cable to run to the utility.

AERIAL SERVICE ENTRANCE

An aerial service to the house is more common than buried service. The utility mounts a step-down transformer on a pole close to the residence, which changes the primary voltage to secondary voltage. (Primary voltage is the high voltage—several thousand volts strong—from the utility. Secondary voltage is the reduced voltage—around 120/240

volts—on the residence side of the transformer.) The service drop, or triplex, which contains the secondary voltage, consists of two insulated hot wires twisted around a bare, stranded, grounded, neutral. The neutral also serves as a support for the triplex by attaching to the house.

To keep the service drop from being hit by traffic or people, the NEC requires that it be 10 ft.—assuming pedestrian traffic only—from the ground at the building (measured from the bottom of the drip loop, not at the point of attachment, because that's normally the lowest point), 12 ft. high over residential property and driveways not subjected to truck traffic, and 18 ft. over the roadway (see the drawing on p. 50). The NEC also says that the service drop cannot cross over the roof of the house unless a clearance of 8 ft. is obtained. And most utility companies don't want it across a roof at all because of possible fire damage. Exceptions are made for steep roofs and significantly high roofs (determined by the utility company and local inspector). Often times it is this height requirement along the service drop that determines where the service drop will attach to the house and in turn where the meter base will be. The service drop can be connected to the gable end (high side) of the house or to the eaves side (low side) of the house.

Wiring Service on the Gable End

When attaching the service drop to the gable end of the house, the utility company terminates the drop by attaching the ground to a porcelain insulator that the workers attach to the house. This is called the point of attachment (see the drawing on p. 51). However, these insulators occasionally pull out if they are only screwed into the siding, so they should be bolted through a supporting member or screwed into solid wood.

The point of attachment is where all three conductors are spliced (by the utility) into the SE cable already installed by the electrician. From this point on, code does not allow service conductors to be spliced (with few exceptions). If the SE cable is exposed to the weather, an exterior-rated cable is required. In addition, you may have to provide

Minimum Clearances over Property

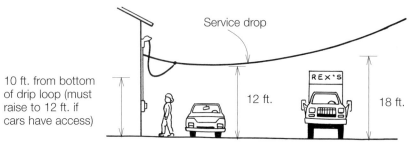

10 ft. from bottom of drip loop (must raise to 12 ft. if cars have access)

Service drop

12 ft.

18 ft.

REX'S

Residential property and driveways

Public streets

From pole to house

NEC calls for minimum clearances on service drop. But ice and heat can cause aerial lines to sag below original construction height. To compensate for sag, construct above minimum requirements.

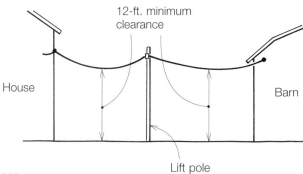

12-ft. minimum clearance

House

Barn

Lift pole

Adding a lift pole

When minimum clearances cannot be maintained across a span, you must add a lift pole.

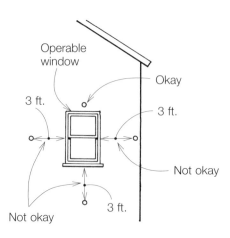

Operable window

Okay

3 ft.

3 ft.

Not okay

Not okay

3 ft.

Around windows

If window is operable, you cannot install individual conductors (a cable such as SEU, which has a jacket, is exempt from rule) within 3 ft. of side and bottom of window, but overhead is okay. If window is not operable, you can get as close as you want. Rule also applies to doors, porches, balconies, ladders, stairs, fire escapes or similar locations.

Point of Attachment

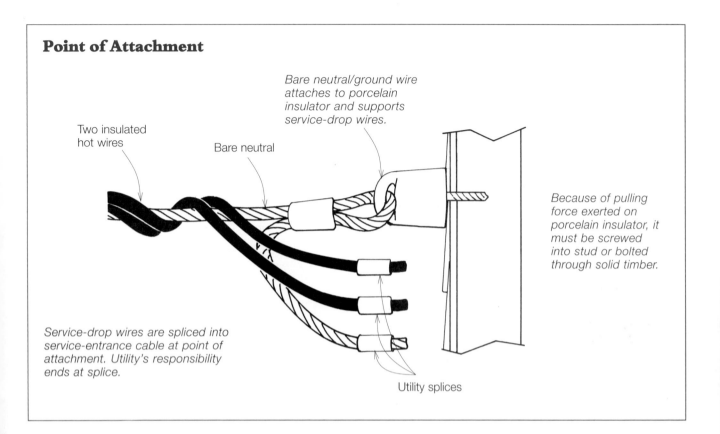

Two insulated hot wires

Bare neutral

Bare neutral/ground wire attaches to porcelain insulator and supports service-drop wires.

Because of pulling force exerted on porcelain insulator, it must be screwed into stud or bolted through solid timber.

Service-drop wires are spliced into service-entrance cable at point of attachment. Utility's responsibility ends at splice.

Utility splices

physical protection for the cable (see section 230-50 of the NEC).

An aerial meter base will have the SE cable entering from the top. When running exposed cable to the service drop, you'll need a water-tight connector that screws into the top, or hub, of the meter base (you will notice threads for the connector on top). If the meter base has no hub, you'll have to install one. Disassemble the connector's threaded top (consisting of a nut, a collar, a rubber grommet and body). Wrap Teflon tape around the connector's bottom threads and screw it into the top of the meter base with a pipe wrench (this could have been done before the base was installed; I just prefer to do it after). Slide the nut and collar and the rubber grommet onto the SE cable and then slide the SE cable down through the connector and into the base.

Strip around 6 in. of the gray outer jacket off one end of the SE cable. Twist all the individual neutral strands together clockwise, to form a braid. After twisting the neutral, it will be about an inch shorter than the two black insulated wires, so trim 1 in. from them. Strip 1 in. of insulation from the two black wires and, if aluminum, coat all three wires with an antioxidant. (Some wire manufacturers recommend wire-brushing the bare wires just before applying the antioxidant.) Insert all three wires under their respective attachments and tighten down (see the drawing on p. 53). Leave no more than ¼ in. of bare conductor showing where the wire slips under the termination. One black wire under either of the outer connections. Neutral to the center. Many localities will require a single ring of white tape wrapped around the neutral wire to mark it as a neutral. Do this on both top and bottom. Though not required by most inspectors, I recommend tightening the wires down with a torque wrench. The inch-pounds required for each nut is listed inside the box.

Typical Gable-end Attachments

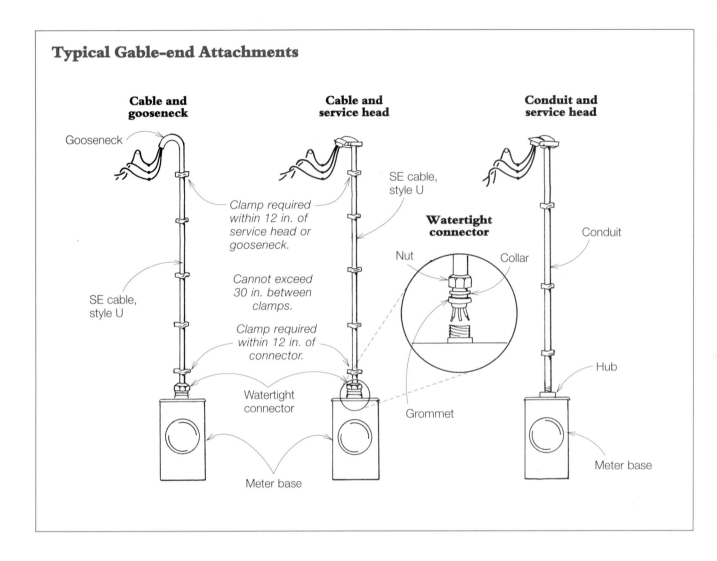

Cable and gooseneck

Gooseneck

Clamp required within 12 in. of service head or gooseneck.

Cannot exceed 30 in. between clamps.

Clamp required within 12 in. of connector.

SE cable, style U

Watertight connector

Meter base

Cable and service head

SE cable, style U

Watertight connector

Nut   Collar

Grommet

Conduit and service head

Conduit

Hub

Meter base

Before climbing your ladder, estimate how much cable it will take to reach the highest point on the wall, and then add at least 4 ft. for the drip loop. That done, cut the cable and start attaching it up the side of the building, and remember that no splices are allowed here. Keep the cable as straight as possible, using a level, a plumb bob or a visual reference. To run the cable, you'll need cable clamps, hex-head screws and a drill with a magnetic bit driver (I prefer to use a cordless drill here so that I don't have to drag a cord up the ladder). Because I'm right-handed, I usually work with the ladder to the left of the cable and with the cable over my right shoulder or in the crook of my arm. The first clamp must be attached within 12 in. of the meter base,

according to the NEC. From there, a clamp must be spaced at least every 30 inches, the last one being within 12 in. of the service head or goose neck (see the drawings above).

If possible, mount the service head or bend the goose neck onto the building about 2 to 3 ft. above the attachment point. However, this may not be possible since you will have no control over where the utility attaches. The NEC wants it to be a minimum of 2 ft. away, but again, the NEC does not control the utility. (The most important thing to remember is that it must be installed so that the drip loop extends several inches below the service head, so water drains *off* rather than *into* the service head and down the SE cable.) Trim the SE cable, strip the outer jacket off

Aerial Meter-Base Connections

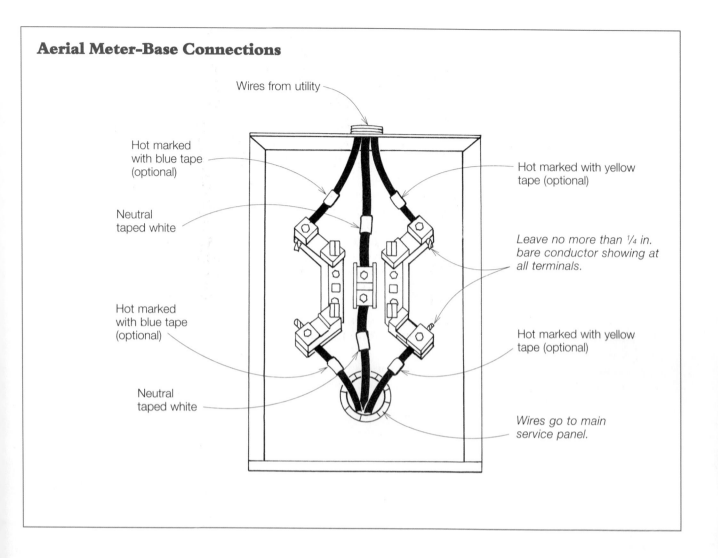

Wires from utility

Hot marked with blue tape (optional)

Neutral taped white

Hot marked with blue tape (optional)

Neutral taped white

Hot marked with yellow tape (optional)

Leave no more than ¼ in. bare conductor showing at all terminals.

Hot marked with yellow tape (optional)

Wires go to main service panel.

and insert it through the service head or bend it into a goose neck. Leave 3 to 4 ft. of cable for the utility.

When finished, the SE cable should have an integral jacket entering the service head and none exiting. Exiting the service head should be two insulated wires and the braided neutral—remember to twist the neutral and add white tape, as you did at the meter-base connection. (In some municipalities, the service head is not required, so be sure to check with your inspector.) If you're doing a goose neck, the cable can bend over to attach to the utility lines. At the meter base tighten the nut of the watertight connector onto the hub body. This connection is rarely as watertight as it needs to be, so it is imperative to put a silicone bead around it.

Wiring Service on the Eave Side

If the meter base is installed on the eave side of the house, the utility attaches the service drop to steel conduit (called a mast) installed by the electrician. The mast is connected to the top of the meter base and is run through the soffit and the roof and capped with a service, or mast, head. The mast provides a stable point of attachment and is a means of running the cables to the meter base.

Locate the meter base where a rafter or main support beam will not need to be cut to install the vertical steel conduit. Attach the base to the wall, then lay out the run. The length of the mast will be determined by the required height needed to get

Making Way for the Service Entrance

The route the utility cables will take to get to the house should be decided upon before construction begins. If the proposed house is a distance away from the main power line, consult with the power company on the best way to get power to the building site. This consultation is important at the planning stage because a secondary line could have a significant voltage drop if it exceeds the maximum distance laid out by the utility. If the house is surrounded by trees, an area will have to be cleared so that electrical lines can be passed to the residence without interference. This also must be coordinated with the utility, and you may have to obtain a "right of way."

A right of way is a legal document that gives the utility permission to install and maintain equipment on your property or someone else's; normally the utility will not install its lines without the form. It is signed and filed in the county courthouse. Once signed, the utility controls the section of property specified in the right of way, not you (but you still pay the taxes).

Utility engineers will visit the site to determine what will have to be done and the costs involved. If the utility has to clear the right of way, many times it will charge you for it—therefore, it is common for the land owner to clear it. This clearance is needed for many reasons. First, maintenance will have to be performed on the line, and the utility will need a legal right to be there to do it. Second, ice storms and strong winds knock down trees, which fall into the lines, so the path must be clear of branches and trees. This is not a local or national code; instead, it's a common-sense requirement of the local utility.

The width of the area to be cleared will depend upon whether primary or secondary power lines will be passing through. An aerial primary line needs at least 40 ft. of clearance. A secondary line normally requires 20 ft. or less, depending on the utility. If the power line must cross land that does not belong to the utility or homeowner, a right of way will have to be obtained from that party, which may not be easy. If the house is in the city, a right of way is sometimes not required, and the problem of getting power to the residence is left up to the utility company.

To keep the cutting to a minimum, you can decide on a buried service entrance, with which you'll need only a path wide enough for equipment to pass through. If placed along the driveway, no additional clearing of trees may be needed. Consult with the utility company for the cost (the homeowner always pays the cost).

from the meter-base top, through the soffit and roof, plus 2 ft. to 3 ft., which is the normal distance above the roof the mast must extend to obtain the required height across the property and the minimum clearances (see the drawing on p. 57). The point of attachment must be 18 in. minimum from the roof, and the mast must be located within 4 ft. of the eaves and less than 6 ft. from the gable end. If the mast must extend more than 4 ft. above the roof, it will need to be braced. Depending on the cable size, the minimum diameter of the mast is 1¼ in., as spelled out by the NEC; I only use and recommend 2-in. diameter galvanized steel pipe (minimum) because of its strength (the pulling pressure created by the weight of the utility cables can be significant).

The pipe is available in standard 10-ft. lengths, and you can get it at electrical supply houses (don't use plumbing pipe; it's not approved for wiring). Try to use one length for the whole run; you'll need the mast to be strong because it also serves as the point of attachment. If you can't avoid splicing the mast with a coupling, make sure that the splice occurs below the soffit, where no significant pressure is

Aerial lines through woods

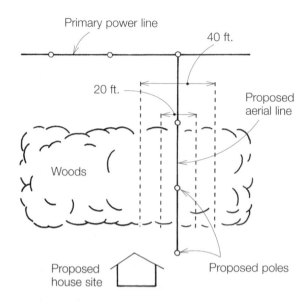

Primary power line

40 ft.

20 ft.

Proposed aerial line

Woods

Proposed house site

Proposed poles

A primary line requires a 40-ft. clearing through woods. A secondary line requires 20 ft.

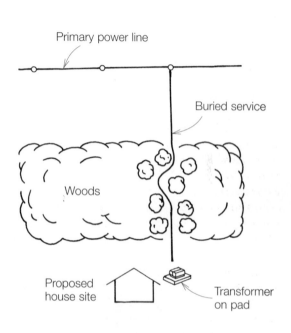

Primary power line

Buried service

Woods

Proposed house site

Transformer on pad

Clearing through woods need only be as wide as a trencher or backhoe. Other equipment can go around trees or boulders.

placed on the mast; if the coupling is above the roofline (an inspector will more than likely reject this placement), the horizontal pulling pressure created by the utility wires could snap the mast at the coupling.

Cut the steel pipe to length with a hacksaw or pipe cutter and ream or otherwise dull the sharp cut edge. Insert the threaded end into the meter base. Use a level or a plumb bob to locate the cut through the soffit, which should be immediately above the threaded top of the meter base. That done, cut

through the soffit using a hole saw (be sure to have your safety glasses on because wood chips will be falling down). I find that a hole saw is best for this job, but if you're careful, you can use a reciprocating saw. After cutting the hole in the soffit, drill a small pilot hole up through the roof as close to center as possible using a long extension bit. Then, from on top of the roof, cut through with the hole saw, using the pilot hole for guidance. This part is easier said than done. If the roof has significant pitch, and most do, the hole will be an oval, not a circle. You have two choices: Use a hole saw and oval out the top and

400-Amp Service Entrance

When running cables larger than 4/0 aluminum, the installation gets tricky because of their thickness. Most of the time, it's easier to run two cables of the same diameter, length and material in the conduit in parallel rather than one large cable. (Never use one large-diameter cable and one small-diameter one because all current would flow through the larger cable because of its lower resistance. Also never mix aluminum and copper.) That is, the electrician wires a 400-amp service with two 4/0 cables going to each leg of the meter base: The two smaller-gauge cables are easier to handle than the one large-gauge cable. In this case it is imperative to mark each phase wire with colored tape. For example, use blue tape to mark the two wires connecting to the meter base's left terminal (mark the cable at the meter base and the service head). The wires connecting to the meter base's right terminal would then have a different colored tape, such as yellow, to identify them (the tape color is dictated by local requirements). The neutral will always be white. Never use green—it is for ground. This way the utility crew would not get the wires mixed up. Another way to prevent any mixup is to use two conduits of the same material, one for each cable group, as opposed to putting all cables into one large conduit (see NEC section 310-4 for conductors in parallel).

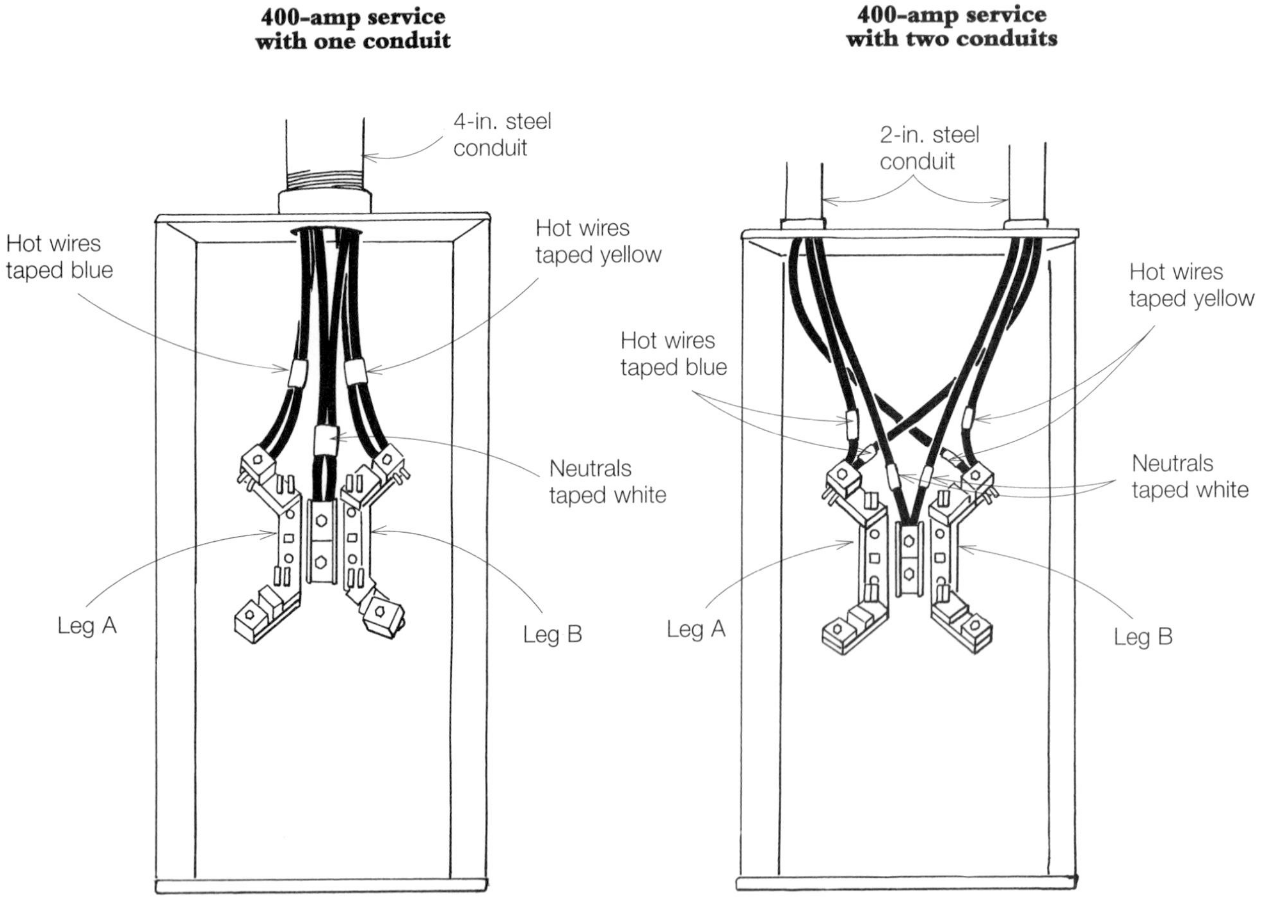

400-amp service with one conduit

4-in. steel conduit

Hot wires taped blue

Hot wires taped yellow

Neutrals taped white

Leg A

Leg B

400-amp service with two conduits

2-in. steel conduit

Hot wires taped blue

Hot wires taped yellow

Neutrals taped white

Leg A

Leg B

Service to the Eave Side

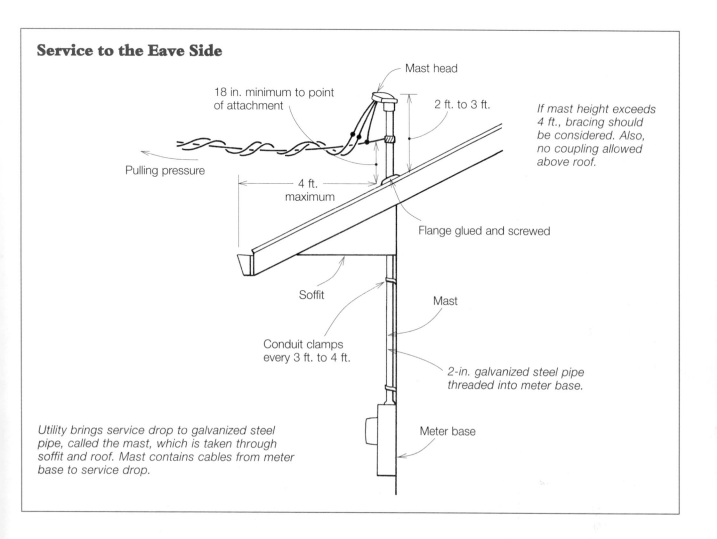

Mast head

18 in. minimum to point of attachment

2 ft. to 3 ft.

If mast height exceeds 4 ft., bracing should be considered. Also, no coupling allowed above roof.

Pulling pressure

4 ft. maximum

Flange glued and screwed

Soffit

Mast

Conduit clamps every 3 ft. to 4 ft.

2-in. galvanized steel pipe threaded into meter base.

Meter base

Utility brings service drop to galvanized steel pipe, called the mast, which is taken through soffit and roof. Mast contains cables from meter base to service drop.

bottom sides. Or cut a piece of PVC pipe at the same pitch as the roof and use it as a template. Or cut the oval hole with a reciprocating saw.

Once the holes are cut, insert the 2-in. dia. pipe from the roof down—threaded end first—through both holes and set it on the meter base's screw threads. Check for plumb. If the mast is not plumb, move the meter base. If you can't move the meter base, move the mast by making additional cuts in the soffit and roof. Once you're sure it's plumb, put pipe dope or Teflon tape on the threads of the mast and screw it into the meter base with a pipe wrench—try not to mar the pipe excessively. Clamp the mast at least every 10 ft. on long runs and within 3 ft. of the meter base, and make sure the clamps are attached to

something secure, such as a stud or beam. During framing, you can install a 2x8 nailer flat against the inside wall (nailed at right angles to the studs) just for this purpose if the outside siding won't be strong enough to hold the mast against the wall. If the mast fits snugly through the roof and soffit, most of the pulling pressure of the utility drop against it will dissipate there.

With the mast attached to the meter base, neatly caulk around the soffit cuts with clear silicon. If the gaps between wood and mast are too big to take caulking, cut a trim piece of wood to fit around the mast. On the roof, caulk between the mast and the roof, install a roof flange over the mast and secure.

Once the meter base and mast are installed, it's time to install the SE cable. Measure from the meter center to the top of the mast, adding enough extra cable to make the loop down to the utility lines. Though it can be done from either side, I normally slip the cable into the mast from the roof, allowing around a foot to stick out of the meter base. I then attach the cable to the meter base according to the directions on p. 51. Remember to apply white tape to any neutral conductor. Once the cable is run, attach the mast head to the mast.

It's possible to have the aerial utility wires hit one side of the house and have the meter on the adjacent side. There are several ways to do this. The most obvious way is to use conduit with an elbow running from one side to the next. This solution is both ugly and expensive. (SE cable may not be able to bend around corners because of its thickness, so individual conductors rated to be used in conduit will have to be installed instead.) Another solution relies on both

luck and the willingness of the inspector. Depending on the situation, the utility wire can terminate close to the house corner and then swing around the corner to attach to the SE cable on the adjacent side. Or you can just run the cable around the corner, which is a very ugly way to do it.

BURIED SERVICE ENTRANCE

Buried is the preferred method of service entrance if the homeowner does not want aerial lines installed or if the meter base is located where an aerial line cannot be installed. For the electrician, this method is easier and less expensive because all he or she has to do is mount the meter base and provide a conduit for the utility to run the cable through. The electrician only has to deal with SE cable running from the meter base to the main service panel. For the homeowner or contractor, the actual expense of buried service will vary. For example, some utilities

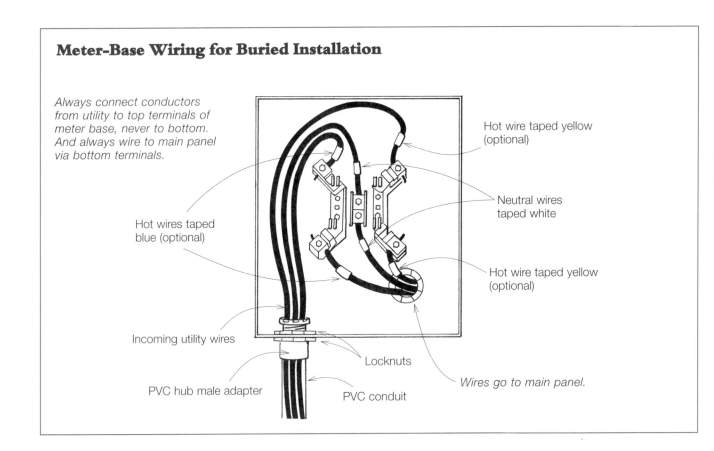

Meter-Base Wiring for Buried Installation

Always connect conductors from utility to top terminals of meter base, never to bottom. And always wire to main panel via bottom terminals.

Hot wire taped yellow (optional)

Neutral wires taped white

Hot wires taped blue (optional)

Hot wire taped yellow (optional)

Incoming utility wires

Locknuts

PVC hub male adapter

PVC conduit

Wires go to main panel.

will give a lower price if a contractor digs the ditch for the service. Regardless, you should contact the water, electrical and gas companies to ask if any of their services run across the path of the ditch.

For a buried service entrance, the utility installs a transformer close to the house. The transformer is usually inside a large, green box mounted on a concrete pad. The purpose of the transformer is the same as the transformer mounted on a utility pole—to change the higher transmission voltage to a lower voltage for the residence. The utility is responsible for getting the wires, called the service lateral, from the transformer to the actual connection at the meter base. The location of this transformer will have to be agreed upon by the homeowner, the contractor and the utility company.

The meter base and connecting SE cable should be in place before the utility arrives to install the buried service. The electrician must provide for the utility crew a section of PVC conduit long enough to reach at least 2 ft. underground with a male adapter, two locknuts and sometimes a long sweep at the bottom. The conduit is attached to the bottom of the meter base with a male hub, two locknuts and, depending on the inspector, a bushing to provide a more rounded surface for the cable to bend against (see the drawing on the facing page). Do not glue on the male adapter or the long sweep because the utility crew may need to cut the pipe to length as they install it. The conduit diameter is normally 2 in. for a 200-amp service, and 3 in. to 4 in. for larger services.

I prefer to have the utility cable enter on the left side of the meter-base bottom instead of the center of the base, as shown in the drawing. This gives me more room to run the cable up the left side to the top connections while leaving ample room to run the exiting SE cable out to the right side. To do this, once the meter base is up (or it can be done beforehand), punch a hole through the bottom left side for the utility. Do not punch too close to the edges of the meter base, or the locknuts won't turn or even fit.

CONNECTING METER BASE TO MAIN PANEL

When designing a residential electrical system, I always try to locate the meter base and the main service panel back to back because it makes connecting the meter base to the main panel easier: Cables can travel through a small piece of conduit connecting the base and the panel.

The SE cables are run through a galvanized nipple (a length of threaded steel conduit) or a length of PVC conduit. You will need an assortment of lengths because you can never really be sure how close the panel is going to be to the base. If the main panel is within the wall cavity, the rear of the panel and the base will almost be touching; if the main panel is outside the wall cavity, the distance could be as much as 6 in. When determining the length of the nipple to use, be aware that around ¾ in. of thread is showing within the base and the panel. The ¾-in. thread allows you to install the locknuts and plastic bushings over the threads. You don't want too much thread showing because it may catch the conductors as they make the bend. A perfect fit would be just enough to get the locknut and the bushing on. You will need four locknuts per nipple—a locknut goes on each side of the meter base and panel. Be sure to get these locknuts *very* tight (see the drawing on p. 60) because the steel nipple is providing a grounding path from the panel to the meter base. If nonmetallic conduit is used, a ground wire should be run to the meter base from the panel's ground bus.

If you decide to use PVC conduit, the distance between the panels will need to be at least far enough to allow two male adapters back to back. Because of the shoulders, or hubs, built into the male adapter fitting, you cannot place a meter base and main panel too close. The PVC male adapters require threaded locknuts, just as the steel nipples do, and may or may not require the end bushings, depending on the inspector and locality. You want just enough thread to get the locknut and end bushing on tightly and no more; a longer length will interfere with the bending of the large conductors.

Back-to-Back Meter Base and Panel

Exterior wall

Exterior siding

Main panel

Meter base screwed into exterior siding

2-in. hole in panel

2-in. hole in meter base

Cut 3-in. hole through exterior siding for steel nipple.

Plastic bushing screwed onto metal threads

Threaded steel nipple

Two pairs of locknuts (grounding locknuts may be required)

When meter base and service panel are installed back to back, a short length of threaded conduit, either galvanized steel or PVC, can be used to pass wires from base to panel.

Before mounting the panel, run the conduit through the meter base into the wall cavity. Place the panel adjacent to the conduit, with its height being adjusted for the main to be around 5 ft. to 6 ft. (6 ft. 7 in. max. to the main breaker) off the finished floor. Mark the conduit location on the gutter of the panel. Then, if there are no prepunched knockouts, use a hole puncher to pop a 2-in. hole for the conduit. Install a locknut on the conduit's threads. Mount the panel—most are made 14½ in. wide so that they can fit between studs—with the conduit threads protruding into the box gutter. Then install the second locknut. If the conduit winds up being too short or too long, replace it.

The wiring from the meter base to main panel always begins at the bottom lugs of the meter base. The most common mistake in wiring a buried meter base is thinking that the top and bottom connections will be reversed as compared to an aerial hookup. These must be made exactly the same as the aerial connections—utility to the top, main panel to the bottom. If the connections get reversed, you will fail inspection. I'll talk about sizing and choosing a main panel in Chapter 4.

Connecting to a Cutoff Panel

If you've got a long distance (each inspector will define this differently—some say 1 ft., others say 6 ft., and so on) to travel from meter base to main panel, you'll have to install a cutoff panel next to the meter base (see p. 73). This cutoff panel, also called a disconnect, provides fault protection for the wires going to the main panel from the meter base. Here's how to install one: From the meter base, run SE cable to the cutoff panel and from there (SER or three insulated cables with ground), to the main panel. Be sure to tape the neutrals white and the insulated grounds green. However, because this situation is common, companies such as Siemens now make meter combinations that have the cutoff panel prewired into the meter base. There is only one box to mount and no extra wiring required.

Some inspectors may allow you to run the cable a few feet to an adjacent wall or basement without requiring a cutoff panel next to the meter base. If

Three Ways to Bring Cable into Basement

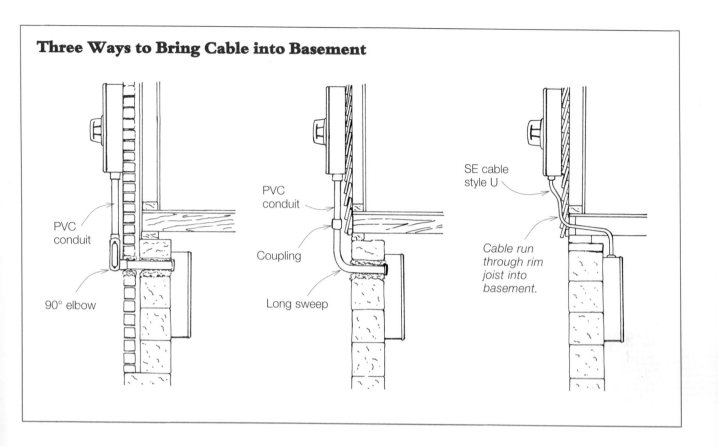

PVC
conduit

PVC
conduit

90° elbow

Coupling

Long sweep

SE cable
style U

Cable run
through rim
joist into
basement.

this is the case you may or may not be required to use conduit (it depends on the type of cable used and local codes). If no conduit is required, standard SER cable can be run to the panel clamped within 12 in. of either box and 30 in. between.

Connecting to Main Panel in Basement

Although I don't recommend it, if the main panel must be located in the basement, there are three ways to get the SE cable to the panel from the meter base (see the drawings above). The first option is to run the SE cable (rated to be outside and in direct sunlight) without conduit right into the basement. However, codes normally don't want you to run cable through concrete without some type of protective sleeve. To get around this, you can run the cable through the siding and rim joist immediately above the concrete and then turn down to the panel—but again, some areas will still require you to protect the cable with conduit.

The most common methods of bringing cable into the basement are to bring PVC conduit immediately out of the bottom of the meter base or cutoff panel just far enough to get into the basement. The PVC normally travels straight down, then through a long sweep or a 90°—immediate-turn—elbow to go through the wall. (Remember to seal the hole around the conduit; otherwise, if the hole is below grade, the basement will flood with the first rain.) Once through, you can either go directly into the main panel, or if you're still above, use an elbow to turn down. This system is expensive, time-consuming, and it's very hard to pull the individual conductors through the conduit, especially the elbows. In addition, water coming in around the meter itself (and sometimes the alleged watertight connector on top of the meter base in aerial installations) will sometimes flow down the cables and into the main service panel, destroying the main breaker. That's why I don't recommend putting the main panel in the basement, unless it's absolutely necessary.

Chapter 4

THE MAIN PANEL

The main panel in the home serves two purposes: It's a master switch that can cut all power at once, and it's a distribution center for all the branch circuits. (Mobile-home panels are a different design than a residence and are not covered here.)

The electrical panel in your residence is one of the most important elements in the electrical system. Many installations fail, regardless of the class of the house, because some installers incorrectly assume that all main panels are designed the same and install the cheapest one they can find. But all main panels are not alike: There are both good designs and bad designs.

ELEMENTS OF THE MAIN PANEL

The main panel (also called a box or load center) has many different sections, with each performing a specific function. The three most important sections of the main panel are the main breaker, the hot buses and the neutral/grounding buses (see the drawing on the facing page). To understand how the panel works, you must first understand what each section does.

Main Breaker

The main breaker, also called the main disconnect switch or just "the main," is normally located at the top of the main panel. All the power that comes into the house goes through this switch. All adults in the home should know what this switch does and where it's located—if there is ever an electrical emergency

Main Service Panel

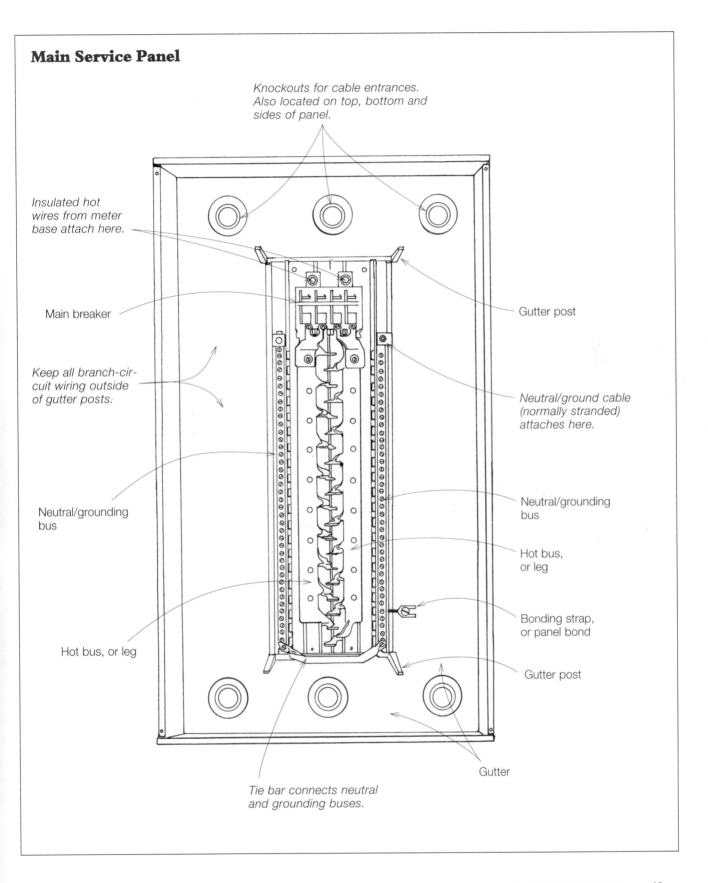

Knockouts for cable entrances. Also located on top, bottom and sides of panel.

Insulated hot wires from meter base attach here.

Main breaker

Keep all branch-circuit wiring outside of gutter posts.

Neutral/grounding bus

Hot bus, or leg

Gutter post

Neutral/ground cable (normally stranded) attaches here.

Neutral/grounding bus

Hot bus, or leg

Bonding strap, or panel bond

Gutter post

Gutter

Tie bar connects neutral and grounding buses.

in the house, throwing this one switch will allow you to cut off *all* power.

The primary purpose of the main breaker is overcurrent protection: The main limits the total amount of current coming into the house so that it cannot exceed the current-carrying capacities of the SE cable and the main panel (overcurrent protection is discussed in detail in Chapter 7). The main senses the amount of current flowing into the panel and automatically disconnects the incoming power from the buses if the amount of current exceeds the amount the breaker is designed for. This amount, in amps, is written on the breaker handle. For example, let's look at a house with a 200-amp main panel. If the branch circuits are pulling 165 amps, and the load suddenly increases to 225 amps, the main will shut off all the power to the house.

The two plastic pieces immediately above the main and below the terminals are called gutter posts. All branch-circuit wiring should remain outside the posts to stay clear of the main breaker terminals (wires cannot cross over the breakers from one side of the panel to the other). The two hex screw terminals immediately above the main are for the SE cables from the meter base.

Hot Bus

Power is transferred from the main breaker to the hot bus, which consists of two copper or aluminum strips, sometimes called legs. The hot bus is located immediately below the main and runs down the center of the panel. Each hot leg has a row of tabs that allow for the insertion of circuit breakers. The bus takes the hot power and distributes it to the circuit breakers.

Each leg is the same voltage, in reference to neutral, but acts as an independent power source (the basics of AC power are covered on pp. 5-6). Simply put, the current in the hot bus comes from alternate sides of the utility transformer to each leg. When measured from leg to leg, the voltages will add up to 240 volts; the voltage for each leg, to ground or neutral, is 120 volts.

Neutral/Grounding Bus

The neutral/grounding bus of the main panel is usually located on opposite sides of the hot buses. Its purpose is to provide a common return point for the current after it has been used by the load. It also provides a ground reference point. Each leg of the neutral/grounding bus is a long aluminum strip with many screws in it. The screws serve as attachment points for the neutral and ground wires (there will be a lot of them). The large hex screw at the top, called the main neutral lug, connects to the neutral coming in from the utility meter base.

Virtually all panel manufacturers have the neutral lug immediately next to the main breaker. This is fine as long as the service entrance comes in from the top. But not every service will enter from the top—many of my entrances come through the panel's gutter. That's why some manufacturers design their panels so that the neutral lug can be removed and reattached anywhere on either neutral/grounding bus—left or right, top or bottom—to make wiring the panel easier. I've found it makes for a much neater installation if I relocate the neutral lug closer to the entry point into the panel. This eliminates running the neutral from the meter all around the inside of the box and leaves much more room in the box for the branch-circuit wires. Removing the lug normally takes a Torx screwdriver.

Even though the neutral is grounded at the utility pole, an additional ground is provided by the installer at the main panel—this is the ground-rod connection and is normally a large-diameter, solid copper wire (4 or 6 gauge). For additional protection, a bonding screw is installed, connecting the neutral/grounding bus to the metal frame of the main panel (see the photo on the facing page). The bonding screw, called the panel bond, places the metal of the panel at ground potential so that it can never become a conductor if a hot wire touches it. In addition, all equipment grounding wires from every receptacle and every appliance, as well as all neutral wires (the white ones), connect to the neutral/grounding bus. (I'll talk more about grounding in Chapter 5.)

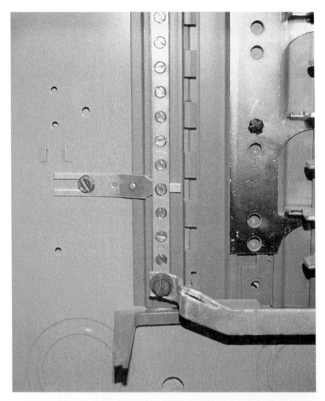

The bonding screw (center of photo), called a panel bond, grounds the panel so that it can never become a conductor if a hot wire touches it.

A well-designed panel will have a bar that connects the two buses as one electrical unit. If the panel with this setup needs to be converted to a subpanel, which has its neutral isolated from the ground, the two can be separated easily by removing the bar. But not every panel has this handy option.

SIZING THE MAIN PANEL

The panel amperage size is dictated by the square footage of the house, along with its electrical load (to learn how to calculate the load, see p. 66). The meter base and the main panel must be sized to handle the same amperage, and the SE cable must be sized for the electrical load as well (for more on sizing cable, see p. 9). Main panels range in size from 100 amps—the smallest—up to 800 amps, depending on the manufacturer. A 200-amp panel is the most commonly installed in a residence in my area of

Virginia; the next most common is a 400-amp panel (this may differ in other parts of the country). I never install a main panel of less than 200 amps, regardless of load calculations, to allow for expansion and electrical upgrades. The higher cost of installing a larger main panel now is insignificant when compared with the cost up a later upgrade. Even if the current design indicates a load of 150 amps, you must think of future loading. It's folly to install a main panel that is fully loaded on the day it is installed. I've lost track of the number of panels I've replaced within the first few years because the installer wasn't thinking of the future. In addition, if the main panel is undersized or is running close to its maximum load, it produces a lot of heat, which will cause the main breaker to trip.

If the load will be over 200 amps, it's best to install two smaller main panels wired in parallel, as opposed to one larger one. For example, if the load will be 300 amps, install a 200-amp panel and a 100-amp panel; if the load is 400 amps, install two 200-amp panels. In this situation, I install all panels next to each other, with SE cable for each coming straight out of a large meter base, which has extra lugs or extra-large lugs provided for the parallel-service wiring required. This type of configuration is better than installing a 300- or 400-amp panel for the following reasons: First, the wiring for such large boxes will be hard to find and install: working with 200-amp cable is almost like trying to bend angle iron—imagine the difficulties of an installation using heavier wiring. Second, finding main breakers for such large boxes is tough, so finding replacements when they go bad may be a larger problem. Third, such large current requirements normally dictate a considerable number of branch circuits that need plenty of space for their hot, neutral and ground wires. The use of two panels gives you more space.

Figure the Number of Branch Circuits

Before purchasing a main panel, you will need to know the number of branch circuits required. Typically, a 200-amp panel has eight to 40 circuits. I normally install a 40-circuit panel. Of these, there are also several designations: I use the one designated 40/40. The first figure refers to the number of tabs

Calculating the Load

The electrical load of a residence is sized according to NEC guidelines. Once you know the load of the house, you can choose the correct-size meter base, cable and main panel for the amperage of the house. In Chapter 9 of the NEC, you'll find formulas for figuring the load of a residence. The one used here is the simplest. For this example, assume a 2,000-sq. ft. house.

1. Calculate the square footage of the house, excluding uninhabited areas such as an unfinished attic, garage and porch, to obtain the general lighting requirements. Multiply this square footage by 3 to obtain the power required: 2,000 x 3 = 6,000 watts, or VA.

2. Add 1,500 VA for each kitchen small-appliance circuit. A minimum of two is required by code, so the number would be 3,000 VA. Then add 1,500 VA for the laundry, excluding the dryer. Total = 4,500 VA.

3. List all appliances to be included and write down their power ratings in volt amps or watts: dishwasher (1,500 VA), water heater (4,500 watts), stove (8,000 watts), clothes dryer (4,000 watts—*Note:* do not use the standard rating of 5,000 watts for the dryer. Use the name-plate rating for the load calculation. Same for stove.), water pump (1,500 VA), garbage disposal (900 VA) and whirlpool tub, etc., but don't include the air conditioner and the heating unit. Let's assume that the total in this case is 29,500 watts or VA.

4. Now add the numbers up: 6,000 + 4,500 + 29,500 = 40,000 watts. Because all loads don't operate at the same time, take the first 10,000 at face value, then take 40% of the rest:

$10,000 + [.4(30,000)] = 10,000 + 12,000 = 22,000$.

5. Take the power of the heating unit or the air conditioner, whichever is largest, and forget the other. In this example it's 20,000 VA and add the 22,000 for the total load: 20,000 + 22,000 = 42,000.

6. Now divide the total wattage by the service voltage: 42,000 watts ÷ 240 volts = 175 amps, which is the total electrical load of the house.

This installation would require a 200-amp main panel, as well as a meter base and cable sized accordingly. If the total load in the home was a bit over 200, you might opt for some gas appliances to lower the current requirements below 200 amps.

available for full-size breakers—one breaker per tab. The second figure refers to the number of circuits available if two-in-one, or dual, breakers are used (see the drawing on the facing page). If the second number is larger than the first, as in a 30/40 panel, this means the panel can hold 30 full-size breakers for 30 circuits. Utilizing the same 30 tabs, 40 circuits can be obtained if the dual breakers are used at specific places within panel. These types of panels normally have the special tabs for the dual breakers located on the bottom of the buses (I'll talk more about circuit breakers in Chapter 7).

The only time I don't use the largest physical panel is when I know I won't need the room. An example is when a panel needs only to house just a few breakers. Let's assume the house service is 300 amps, of which the heat-pump system will take something less than 100 amps. I use two main panels side by side, a 200-amp and a 100-amp panel, both wired directly into a 400-amp meter base. I put the heat-pump breakers in the 100-amp panel, leaving the 200-amp panel for the branch circuits and other house appliances. The 100-amp panel will have two double-pole breakers—a 30-amp for the outside unit and a 60-amp breaker for the inside unit (amperage

The Right Tab for the Breaker

Full-size breakers fit both standard-tab or slotted-tab buses.

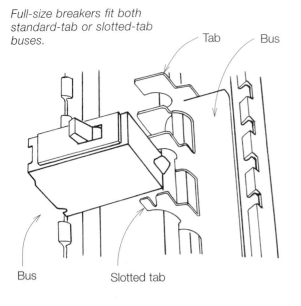

Tab   Bus

Bus   Slotted tab

Two-in-one, or dual, breakers fit only tabs with slots.

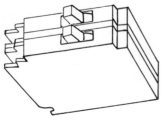

of the breaker depends on the size of the unit and its electric backup). This is a high-current system but only needs four tabs in the bus: two for each double-pole breaker. If two heat-pump systems are being installed, double the amount of tabs needed to eight. If you absolutely, positively know you're not going to need much breaker space, you don't need a large panel.

CHOOSING A MAIN PANEL

A 200-amp panel is the most common panel used in today's homes. But as I have mentioned, all panels are not the same. Some designs are great; others I suggest only to my enemies. I only buy well-designed panels made by manufacturers that watch out for me, the electrician. I avoid panels with illogical designs that take longer to install, make the box look messy and can be dangerous to the electrician. For instance, I've seen a panel whose breakers cover the neutral bus when full-size breakers are installed. The inspector can't even see if there is a loose connection. And even if there were, you would have to remove the breaker just to tighten it. Another manufacturer makes a panel in which you have to remove several breakers just to remove one. The bottom line is, buy a quality panel, not a cheap one.

Look for Lots of Wiring Room

I want an intelligently designed panel, one that makes installation and operation safe and simple. When choosing a main panel, look for the largest allowable, in both number of circuits and amount of space. For instance, with a 200-amp load, I normally choose a full-size 40/40 panel.

I hate working in panels that are so overcrowded that you have trouble even installing the cover, or lid. That's why I look for a panel with lots of wiring room. A physically large panel—the largest is a 40/40—will avoid the "bird's nest" that you see too often within the panel. I've seen bird's nests so bad that I couldn't put the cover on the panel. In addition, a 40/40 panel allows the use of all full-size breakers.

Make Sure Neutrals are Placed Logically

Logical placement of neutrals in the panel is a must. I think it's best to have two neutral buses, one on each side of the hot buses, next to the breakers. This way both the hot and neutral wires are always cut the same length, which results not only in a neater box but also in labor savings. Placing the neutral bus in this way makes perfect sense to me, but many manufacturers don't do it. Many just give you one neutral, not two, and stick it off to one side, which

What to Look For in a Main

Wide handle allows you to get several fingers on it.

Narrow handle is hard to grab and operate.

A main that turns on/off vertically is easy to operate.

A main that turns on/off sideways is hard to operate.

Good design

Poor design

The poorly designed neutral on this panel is buried under, and adjacent to, the hot buses.

On this panel, the poorly designed neutral is located off to one side—stacked, and away from the breakers—which forces the electrician to cut the neutral wires longer than the hot wires, increasing labor time dramatically (especially when dealing with many circuits).

doesn't make sense because the neutral wire must be twice as long as the hot wires and gets in the way of other cables. I also hate it when the manufacturer stacks the neutrals on top of each other so that you have to remove one wire to get to another (see the photo at left). One well-known manufacturer stacks its neutrals immediately below the hot-wire connection to the main breaker (see the photo above). If you slipped with a screwdriver or a meter probe—ZAP! Sometimes I think manufacturers are trying to kill me.

Even if the neutral buses are in the right location, they may be too short. Short neutral buses crowd the neutrals into one spot, creating a bird's nest, and making you double up—install two neutrals under one screw—because there's not enough room. Many times you have to go out and buy an additional neutral bus and install it in the panel yourself.

Look for Quality Hardware

Make sure the panel has a main with a full-size handle—one you can get four fingers on. Some mains have handles smaller than my little finger, which are very hard to turn on and off. I've had to make many a service call simply to turn power on or off for a customer because the main handle was too small for him or her to grab. In addition, look for a main that turns on/off vertically: It's much easier to pull down to turn the main off than to try and pull it to the side (see the drawing on the facing page).

Try to choose a panel with tabs on the panel cover, which keep it from falling when the last screw holding the cover to the panel is withdrawn. Many an electrician has had his toes smashed or cut because of the cover falling. The tabs also hold the cover in place as you install the screws. Before I found a cover with tabs, I would shoot a couple long screws through the top of the cover. The screws catch on top of the box and keep the cover from falling straight down and amputating my toes.

Pick a panel in which the breakers can be installed and removed easily. This sounds obvious, but believe it or not, some manufacturers make you remove an entire row of breakers before you can install or remove just one.

Look for a neutral/grounding system that easily converts a standard main panel to a subpanel. This keeps you from having to stock two different panels. A tie bar at the bottom of the neutral buses connects the two buses. The two can be separated easily by removing the bar, which will separate ground and neutral. Make sure the panel has a removable neutral lug, which allows the lug to be moved to other locations on the bus, and hot and

neutral lugs with heavy-duty hex screws that allow for hard tightening. The old-fashioned screwdriver tightening system simply cannot get the terminal tight enough. In addition, to make tightening the screw easier, the hex screw should be a large-diameter hex, as opposed to a small diameter. A small-diameter hex wrench simply cannot get the screw tight enough without ruining the wrench.

BALANCING THE LOAD

In theory, and as much as practical, the load needs to be balanced as it leaves the panel and enters the service-entrance wires. A balanced load is one that is wired in the panel so that the current on one 240-volt leg is equal to the current on the other, so the current cancels out in the neutral. Balancing requires you to analyze the circuits to determine the load they are going to serve, and then you must employ some common sense.

Balancing lowers the heat buildup in the wire terminals and allows you to put more load on the panel. Let's assume you have a 200-amp panel and main breaker. If you wired the load so that 300 amps of current was being pulled through only one leg, the main would trip. However, if you split, or balanced, the load, the current would be reduced to 150 amps on each leg—well within the 200-amp rating of the main.

In most main panels, the tabs on each leg of the hot bus are arranged in opposite phases, so balancing occurs automatically. As single-pole breakers (those that use only one hot leg of the panel) are installed down one side of the panel, they automatically get put on alternate phases because of the tab/bus design. If a double-pole breaker (one that uses both hot legs of the panel) is installed, its current will automatically balance because the breaker, by design, fits on opposite phases. (I'll talk more about single- and double-pole circuit breakers in Chapter 7.)

Inside the house, if two loads of any kind will be on at the same time, put them on opposite phases. For example, if two circuits—one with a freezer pulling

5 amps and one with a refrigerator pulling 5 amps—are put on the same phase, their current in both the neutral and on the service-entrance hot leg will add up to 10 amps. If the same circuits are put on opposite phases, 5 amps will flow through each hot leg and will cancel out to 0 amps in the neutral. This is why the neutral is allowed to be smaller than the two hot wires. Even if the neutral load doesn't cancel, it will be reduced by the amount of the smallest load: If one load pulls 10 amps, the other 6, the resultant current flow through the service-entrance neutral will be 4 amps if the loads are on opposite phases—16 if they're not (see the drawings on the facing page).

If you're building a shop with a 240-volt feed, and two 120-volt machines are going to be running at once, put them on opposite phases so that the current will cancel in the neutral wire leading to the shop, as well as in the service-entrance neutral cable. If the shop only has a 120-volt feed, there is nothing you can do. It takes two opposing currents to cancel. With only one current, nothing will cancel.

SUBPANELS

Subpanels are used where the main panel is a considerable distance (normally 6 ft. but depends on locality) away from a major portion of the wiring. A subpanel makes wiring this situation easier because it allows you to run just one cable to the subpanel, and then feed the appliances and branch circuits on that feeder off the subpanel. I would think twice about including a subpanel in the plan, because it may not be as convenient as it sounds. The extra expense of the materials required to wire a subpanel may not outweigh cost of labor to run multiple wires a long distance.

A subpanel requires a four-conductor cable (SE, type R), with two insulated hot wires, an insulated neutral and a grounding wire that connects all the way back to the main panel. Expect to pay around $3 per ft. for a cable rated to 100 amps, and it may not be easy to find. If your wire run is long, this could be an

expensive proposition. You also have to buy the panel. It would be more expensive, materialwise, to install a subpanel in a standard residence than it would be to run 100, 12-gauge wires to the appliances. Savings, if any, will be in the labor. It obviously would take less labor to run a single cable than it would to run 100 wires.

A subpanel is called a lugs-only panel because it has no main breaker. Lugs for wire attachment are in its place. You can add a main by simply pressing in an appropriate-size breaker into the tabs of the subpanel if you'd like. The incoming four-conductor—three insulated, one ground—wiring from the main panel is then brought in to power the subpanel. A main breaker in the subpanel, though not necessary, gives you a convenient main disconnect for the subpanel. The result will be two disconnects: one at the main panel and one at the subpanel.

Wiring a Subpanel

Once you've determined that you need a subpanel, installing one is rather simple. Buy a subpanel, or lugs-only panel, for the correct amperage size. You can remove or ignore the lugs and install a breaker if you want. Be sure the subpanel has two bus bars: one electrically isolated on plastic feet for neutrals, the other bonded to the frame for grounds. The neutral bus must not be connected to the panel frame—only the grounding bus can do that in a subpanel. This is to prevent neutral current from flowing through the grounding system on the load side of the main breaker (which is a code violation). Doing so would place current through the metal conduit and ductwork, creating a shock and fire hazard. I once did a service call where there was so much current flowing through the loose ductwork connections that the subpanel looked like it had Christmas tree lights inside it. The arcing and burning noises kept the homeowners from sleeping at night. Any Siemens main panel can be used for a subpanel by simply removing the bottom bar that connect the two buses (see the photo on p. 72). Once removed, neither bus is connected to anything (these are called

Balancing the Load

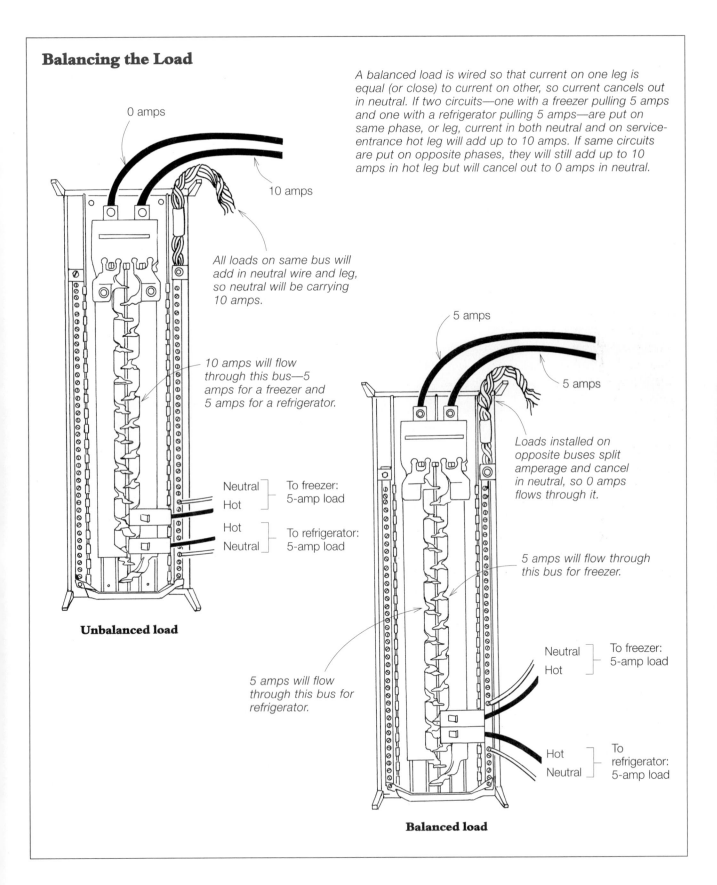

A balanced load is wired so that current on one leg is equal (or close) to current on other, so current cancels out in neutral. If two circuits—one with a freezer pulling 5 amps and one with a refrigerator pulling 5 amps—are put on same phase, or leg, current in both neutral and on service-entrance hot leg will add up to 10 amps. If same circuits are put on opposite phases, they will still add up to 10 amps in hot leg but will cancel out to 0 amps in neutral.

0 amps

10 amps

All loads on same bus will add in neutral wire and leg, so neutral will be carrying 10 amps.

10 amps will flow through this bus—5 amps for a freezer and 5 amps for a refrigerator.

Neutral
Hot
⎤ To freezer:
⎦ 5-amp load

Hot
Neutral
⎤ To refrigerator:
⎦ 5-amp load

Unbalanced load

5 amps will flow through this bus for refrigerator.

5 amps

5 amps

Loads installed on opposite buses split amperage and cancel in neutral, so 0 amps flows through it.

5 amps will flow through this bus for freezer.

Neutral
Hot
⎤ To freezer:
⎦ 5-amp load

Hot
Neutral
⎤ To
⎦ refrigerator:
5-amp load

Balanced load

This Siemens panel easily converts to a subpanel simply by removing the tie bar that connects the neutral and grounding buses.

floating buses). Pick one for the grounding bus, and one for the neutral.

The box location is determined by the area most densely populated with branch circuits, often the kitchen. Simply mount the box like any other. The connecting cable will need to be SE type R, with three insulated conductors and a ground. You can also run individual conductors through conduit if you prefer, but this method can be expensive and time-consuming.

Connect the two hot wires to the appropriate size double-pole breaker in the main panel, and connect the opposite ends in the subpanel to the lugs or another breaker (use the same size breaker as the one the wires are connected to in the main panel). Connect the neutral to the neutral bus in the subpanel and to the neutral/grounding bus of the main panel. Connect the grounding wire to the grounding bus of the subpanel and to the neutral/grounding bus of the main panel (see the drawing below).

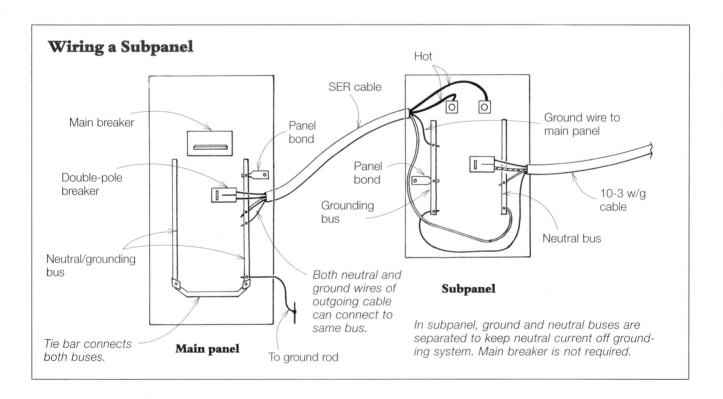

Wiring a Subpanel

Main breaker

Double-pole breaker

Neutral/grounding bus

Tie bar connects both buses.

Main panel

Both neutral and ground wires of outgoing cable can connect to same bus.

To ground rod

Panel bond

SER cable

Hot

Panel bond

Grounding bus

Ground wire to main panel

10-3 w/g cable

Neutral bus

Subpanel

In subpanel, ground and neutral buses are separated to keep neutral current off grounding system. Main breaker is not required.

The appliance is wired with three-conductor cable with ground. Receptacles and switches are treated no differently than if they were wired from a main panel, except for the connections in the subpanel, where the neutrals and grounds are separated.

Converting a Main Panel to a Subpanel

When a cutoff, or disconnect, panel is needed to provide fault protection for a long run of SE cable to the main panel, the main panel has to be wired as a subpanel. The neutral bus will have to be disconnected from the grounding bus, and the panel bond and the grounding wire to the ground rods must be on the grounding-bus side. All white neutral wires connect to the floating neutral bus, and all bare and green grounding wires connect to the grounding bus. Panels made by Siemens have a removable bar at the bottom of the neutral/ grounding bus. Once the bar is removed, one bus can be the neutral bus, and the other can be the grounding bus (see the drawing below). It makes no difference which. Some main panels will not allow you to do this.

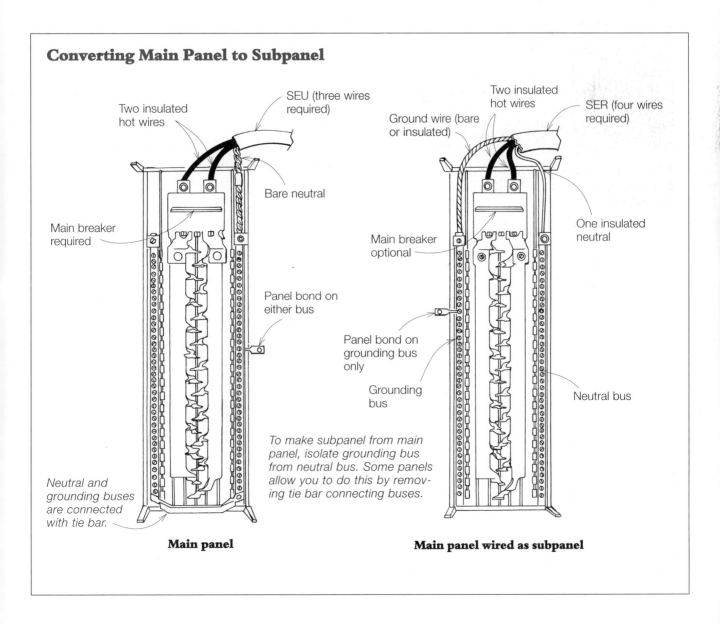

Converting Main Panel to Subpanel

Two insulated hot wires

SEU (three wires required)

Bare neutral

Main breaker required

Panel bond on either bus

Neutral and grounding buses are connected with tie bar.

Main panel

Two insulated hot wires

SER (four wires required)

Ground wire (bare or insulated)

One insulated neutral

Main breaker optional

Panel bond on grounding bus only

Grounding bus

Neutral bus

To make subpanel from main panel, isolate grounding bus from neutral bus. Some panels allow you to do this by removing tie bar connecting buses.

Main panel wired as subpanel

Chapter 5

GROUNDING

GROUNDING OFFERS
PROTECTION

THE GROUNDING
SYSTEM

WHAT TO GROUND

Grounding is at the foundation of the electrical system. An improperly grounded electrical system is a hazard that can damage or destroy appliances and even kill people who use them or come in direct contact with them. It is imperative that residences, especially those with computers, answering machines and other high-tech sensitive equipment, have an extremely good grounding system. Otherwise, these expensive machines are in jeopardy.

In this chapter I will explain how to install a proper grounding system. And I will emphasize that care must be taken when installing the system because one mistake can cause the entire system to fail.

GROUNDING OFFERS PROTECTION

Grounding is the process of connecting all noncurrent-carrying conductors in the electrical system and then tying them to the earth via a ground-rod system. (A noncurrent-carrying conductor is a material that *can* conduct electricity but normally does not, such as the metal frame of a washer or dryer or the metal housing of an electric drill.) A grounding system protects against ground faults (a short circuit that occurs when a hot wire touches a ground or noncurrent-carrying conductor), induced voltages and voltage surges in the electrical system.

Ground Faults

If a hot wire accidentally touches the grounded frame of a tool or appliance and energizes it, a ground fault has occurred. This is one of the more common of tool/appliance malfunctions. The equipment grounding conductors provide a return

Ground Fault on Grounded Appliance

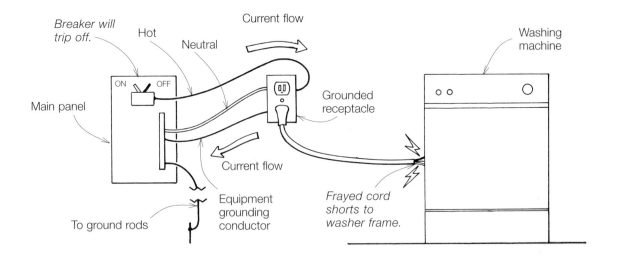

Once frame of washer is energized, ground-fault current flows back to main panel via equipment grounding conductor. No current flows to ground rods; it stays on wiring and trips breaker.

Ground Fault on Ungrounded Appliance

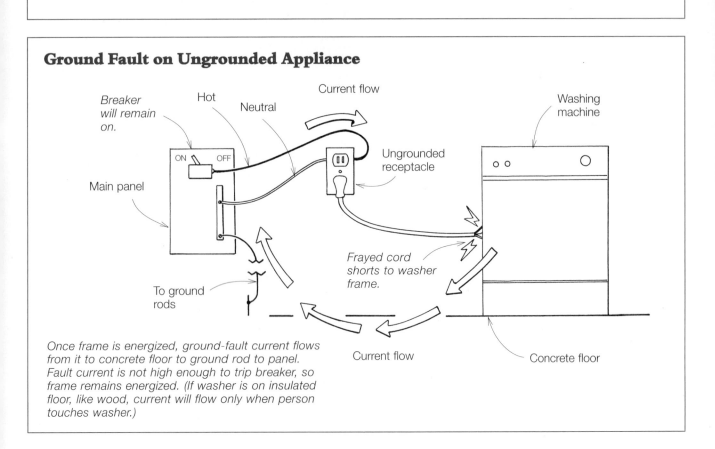

Once frame is energized, ground-fault current flows from it to concrete floor to ground rod to panel. Fault current is not high enough to trip breaker, so frame remains energized. (If washer is on insulated floor, like wood, current will flow only when person touches washer.)

path to the main panel's neutral/grounding bus for ground-fault current flow (the earth is not in the current flow path).

For example, let's assume that the cord of a washing machine has frayed, and the hot wire is touching the metal frame, resulting in a ground fault (see the top drawing on p. 75). Once the frame is energized, the fault current will flow back to the main panel via the equipment grounding conductor, and the circuit breaker will trip off (I'll explain how a circuit breaker works in Chapter 7). In this situation, the fault current does not flow to earth ground to dissipate. It remains on the low-resistance wiring to the breaker, allowing the breaker to trip off.

If an equipment grounding conductor is not present or has a bad connection somewhere along its path, the breaker cannot trip because there is no way for the fault current to get back to the main panel to complete the circuit. The frame of the clothes washer will remain energized, and a person who touches it can receive a serious shock. What's happening here is that the person's body becomes an electrical conductor. And the less the person is insulated from the ground (standing barefoot on a damp basement floor), the greater the current flow through the body.

A ground fault doesn't have to be direct, either. Though the initial fault or short to appliance may be direct, the routing back to the panel may be high resistance. The earth may be a conductor, but it is normally not good enough to allow sufficient fault current to trip the breaker (see the bottom drawing on p. 75). Although the current may not be high enough to trip a breaker, it could still hurt or kill you, or damage the appliance.

Unwanted Voltage and Current

The earth grounding system attempts to place all noncurrent-carrying conductors at 0-volt potential by bleeding off any stray currents into the earth through the grounding conductors. Without grounding conductors to direct the unwanted current away, and an earth ground system to dissipate it, the stray voltage can build up to damaging potential.

Stray current resulting from induced voltage can come from any current-carrying source (current through one wire can induce current into another, for example). Induced voltage also comes from any continuous voltage source, such as an overhead high-voltage line. The power line could induce voltage and stray current into a home with a metal roof or metal framing, where a good grounding system does not exist. To avoid this type of voltage buildup, the metal must be well-grounded into a system of ground rods (I'll talk more about ground rods later in this chapter).

Surge current comes primarily from lightning. The magnetic field created by a lightning strike, even if it's over a mile away, can induce voltage into any current-carrying conductor (like house wiring) or noncurrent-carrying conductor (like metal water pipes and appliance frames), producing a short-lived high-voltage spike. If the induced voltage and current find no path from the appliance to ground, it could go through sensitive electronic equipment to find its own, or simply arc through the air. The resulting shock to humans can vary from a minor jolt to electrocution.

Surge Arresters

Surge arresters suppress voltage surges (that is why they're also called suppressors) by directing excess voltages and currents into the grounding system, which provides a path and location for the current to dissipate. But neither an expensive secondary surge arrester in the main panel nor a cheaper point-of-use arrester will work properly without a high-quality grounding system.

Different surge arresters work different ways. I use the Tytewadd surge arrester for three reasons: (1) It mounts inside the main panel, not outside (surge arresters can explode); (2) it indicates when it's bad; (3) it works. I've used many different types of surge arresters, and none comes close to this one. Here's how the Tytewadd works. If lightning induces a surge voltage into the power line, the surge will charge into your house through the service entrance. (The surge could be thousands of volts strong.) With a good grounding system, a good secondary surge arrester, such as the Tytewadd, hard-wired at the panel will clip the excess voltage off the line and throw it into the low-resistance grounding system (see the drawing below). A point-of-use plug-in type sends excess voltage through the equipment grounding conductor back to the main panel and into the ground. If a poor grounding system is in use, the surge voltage will increase at the point of highest resistance, such as a loose or corroded connection, and could arc over and destroy any appliance in the area.

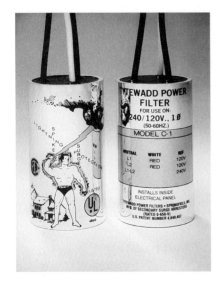

Surge arresters take power surges off the line and dissipate excess voltage and current into the grounding system. Both of these units are made by Tytewadd: one is attached at the panel, and the other is a plug-in type.

How Surge Arresters Work

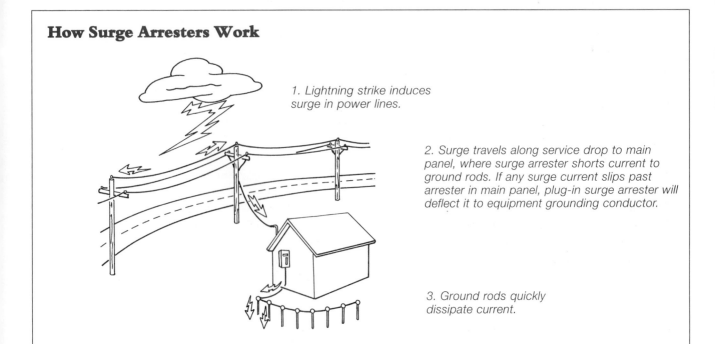

1. Lightning strike induces surge in power lines.

2. Surge travels along service drop to main panel, where surge arrester shorts current to ground rods. If any surge current slips past arrester in main panel, plug-in surge arrester will deflect it to equipment grounding conductor.

3. Ground rods quickly dissipate current.

THE GROUNDING SYSTEM

The goal of the grounding system is to tie all noncurrent-carrying conductors together and then place them at earth ground potential (0 volts). But all grounding systems have a common enemy: resistance. As I mentioned in Chapter 1, resistance is the opposition to the flow of current—in this case it's grounding current, which could be stray current from a ground fault or surge current. As resistance to the grounding current increases, the voltage in the grounding system will also increase. If this voltage gets high enough, it could damage appliances. For example, an illegal, loose or corroded clamp will provide a poor connection to the rod and may have a very high resistance (100 ohms or more). A typical lightning pulse could induce several thousand amps across the clamp, which could develop several thousand volts as well. This high voltage is felt throughout the grounding system and will arc over through sensitive electronic equipment to get to ground if it finds a lower resistance path. For some unknown reason, this normally occurs through the most expensive electronic gear, such as a color TV.

Your objective in designing a grounding system should be to provide as little resistance as possible to allow current to flow easily to the ground. A good grounding system starts with the actual earth connection itself: the ground rods. Though other devices may be used in lieu of ground rods, I do not recommend them. Ground rods don't have to be used in all grounding systems. (The NEC does allow foundation rebar to be used in some situations, but don't do it. Surges could be high enough to turn any moisture within the concrete to steam and crack the foundation.)

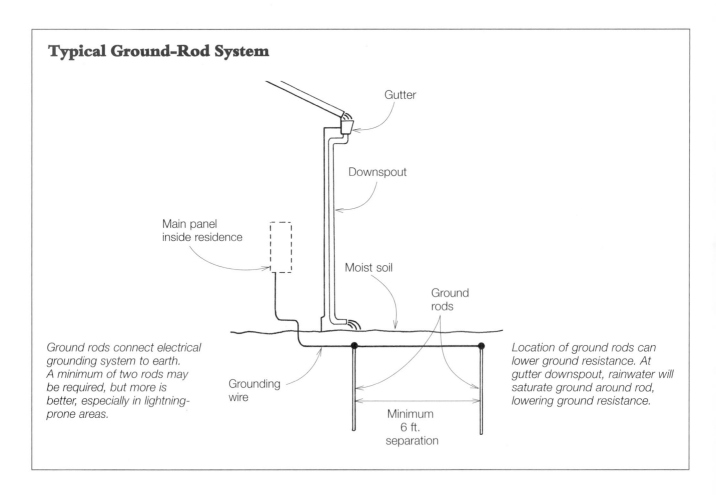

Typical Ground-Rod System

Gutter

Downspout

Main panel inside residence

Moist soil

Ground rods

Ground rods connect electrical grounding system to earth. A minimum of two rods may be required, but more is better, especially in lightning-prone areas.

Grounding wire

Minimum 6 ft. separation

Location of ground rods can lower ground resistance. At gutter downspout, rainwater will saturate ground around rod, lowering ground resistance.

Lowering Ground Resistance in Dry or Rocky Areas

In areas where the soil has a high resistivity (where the ground is dry or very rocky), you can lower the resistance of the ground by pouring a ground enhancement material around the driven ground rod. In essence, the material artificially increases the diameter of the ground rod. Be sure to use a brand that does not contaminate the ground water (read the label). I prefer a product called Gem—available at some electrical supply houses.

Here's how to use it: Dig a hole about 3 in. to 6 in. in diameter (use a post-hole digger or an auger). Drive the ground rod in the center of the hole and then pour the ground enhancement material around the rod. Once done, you'll have a ground rod the diameter of the hole. Ground enhancement materials also work well in rocky areas, where rods must be placed flat in trenches as opposed to being driven. In this installation, the enhancement material is poured above and below the rod, giving a very large surface contact area with the soil.

Installing the Ground Rods

Ground rods (they're also called grounding electrodes) are driven into the earth close to the main panel or utility meter base. They connect the grounding system to the earth, dissipating the stray voltages and currents on the electrical system within the house. But the rods could also work in reverse, providing the pickup point in the earth for the return current path when a ground fault occurs somewhere in the yard. For example, if you're using a drill in the yard and get a ground fault, current will flow down through you, through the wet grass and into the ground rods, which would allow the current to return to the panel. The latter is not an intended purpose of the rods—the earth's resistance will normally attenuate the current well below the level needed to trip a breaker. However, for troubleshooting purposes, one should know that the rods will conduct electricity both ways.

Size and material Ground rods must be at least 8 ft. long (10 ft. is better). A longer rod has a better chance of lowering the grounding resistance by penetrating into deeper, moister soil. However, do not try to add a second rod on top of another to get deeper into the ground: It does not significantly lower the ground resistance. It's better to drive a second rod a few feet away from the first (NEC requires minimum spacing of 6 ft.).

A larger diameter ground rod provides less grounding resistance than a smaller diameter rod. The minimum diameter allowed by the NEC is ⅝ in. A rod with a diameter of ¾ in. is obviously better, and most wholesale electrical supply houses carry them.

Most ground rods are made from copper-clad steel or galvanized steel (both coatings prevent corrosion). Galvanized pipe ¾ in. or larger in diameter can be used in lieu of solid rods, but I don't recommend it. Ground-rod clamps that are rated for use outside may not fit over galvanized pipe, and you'll probably substitute some illegal clamp that will corrode. However, if you find a clamp that is rated for outside installation and is large enough to fit the pipe, such an installation can be considered.

Locating and driving the rods The number of ground rods you install is very important because it will directly affect how good a ground you will have. The more rods that you install, the lower the ground resistance. Code demands two rods, unless you can prove a resistance of 25 ohms or less (many homes only have one rod, which is not enough in most cases). I install eight rods if lightning is a problem; otherwise, two should be sufficient.

Though they should be driven somewhere reasonably close to the main panel, it is more important to locate the ground rods where the ground is moist and can provide a low ground resistance. A good location close to the house is normally around the gutter downspouts and the drip edge of the house, as shown in the drawing on p. 78. But if lightning is a problem in your area, you'll want to make extra-sure that you've got a really good, low-resistance earth ground. To ensure a proper ground, you may have to run the grounding wire more than 100 ft. to get to a creek bed, an abandoned pond or wooded valley. If the inspector objects because the rods are not close enough to the main panel, simply install a couple ground rods at the meter base to satisfy his requirements, then continue running the grounding wire to where you need it to get the necessary ground.

To install a rod, dig a small hole about 6 in. to 8 in. deep, using a post-hole digger. Drive the rod, leaving about 4 in. above the bottom of the hole. The easiest way to drive a ground rod is with a hammer drill (you should be able to rent one). The most common way to drive a rod, however, is with a sledgehammer. The problem with using a sledgehammer is that the rod swings around as you try to hit it. To help hold the rod steady, I drill a 1-in. hole through a scrap 2x and put it over the rod. One person holds the board to steady the rod as the other hammers it. Try not to "mushroom" the rod, or you'll have to file it to fit the clamp over.

If you're unable to drive the rod all the way in because you encounter solid rock, you've got three options (see the drawings below): You can drive the rod at an angle up to 45° from vertical or simply lay the rod in a trench (minimum 2½ ft. from the surface, or grade). Another option is to dig a trench and drive the rods at 45° until you hit rock. At that point, bend the rod flat into the trench.

Running the Ground Wire

Also called the grounding electrode conductor, the ground wire is the large, bare copper wire that connects the ground-rod system to the grounding bus within the main panel. The gauge of the ground wire needed is dictated by the NEC and depends on

Driving Ground Rods in Rocky Areas

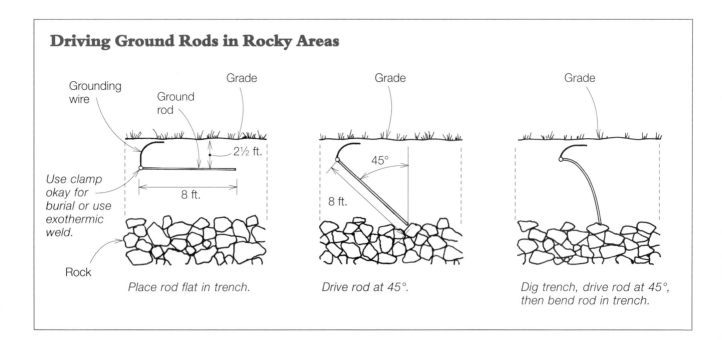

Place rod flat in trench.

Drive rod at 45°.

Dig trench, drive rod at 45°, then bend rod in trench.

Sizing Ground Wire

Size of largest SE conductor		Size of grounding electrode conductor	
Copper	Aluminum or copper-clad aluminum	Copper	Aluminum or copper-clad aluminum
2 or smaller	1/0 or smaller	8	6
1 or 1/0	2/0 or 3/0	6	4
2/0 or 3/0	4/0 or 250 kcmil	4	2
Over 3/0 to 350 kcmil	Over 250 kcmil to 500 kcmil	2	1/0

Where connected to made electrodes, such as ground rods, the grounding electrode conductor does not have to be larger than 6-gauge copper. However, for any other use, such as jumpers around water heaters and water meters and ground connection from water pipes to panel, the conductor must be full size (from chart). For 200-amp service using 4/0 wire, full size would be 4-gauge copper.

the size and type of conductors in the service (see the chart above). In a typical ground-rod system, 6-gauge copper is sufficient. But if I have a long run, say over 20 ft., the resistance in the wire is increased. So I compensate for the increased resistance by using thicker, 4-gauge wire.

Do not cut the wire anywhere along the run—it must be continuous from beginning to end, regardless of the number of ground rods you plan to install. The inspector may make an exception if the splice can be done with an irreversible, compression-type connection designed and listed for that purpose and an exothermic welding process (for more on exothermic welding, see pp. 82-83). The wire must also be buried along its run to protect it (minimum 6 in.), and it should enter the wall cavity as soon as possible for its own physical protection. If the wire being run along the outside of the residence is not subject to damage, it can be run without conduit as long as it is stapled flat against the wall approximately every 2 ft. (the wire can be painted to match the building's color scheme); if the wire could

be easily damaged, run it through conduit. All wire smaller than 6-gauge copper must be run through conduit for protection, but I don't ever use a ground wire smaller than 6 gauge.

If the residence has two main panels, the ground wire can be taken to either one and then connected to the second panel. However, the inspector may not allow splices in the bus bar or the grounding wire, so the wire should be looped through one of the holes in the grounding bus of the first panel and run to the second. Be sure that the set screws clamping the grounding wire down to the grounding buses are tight.

Use the right clamp Clamps attach the ground wire to the ground rod. The attachment won't be approved unless the clamp is listed for outdoor use. The most common clamp is called an acorn clamp. But all clamps add resistance to the grounding system, so I prefer to use an exothermic weld, which adds no resistance (see pp. 82-83).

Here are a few tips to help you pass inspection:

1. Do not use a pipe clamp that is not rated for outside use but is only rated for inside water-line attachment.

2. Do not just wrap the wire around the rod.

3. Do not use hose clamps to attach the ground wire to the ground rod.

4. Soldering the wire onto the rod to lower the resistance is a good idea, but you must also use a listed clamp for added holding power in case a surge current melts the solder.

Grounding the Main Panel

The main panel will always have a panel bond, which is simply a wire or metal strap that connects the panel into the grounding system (see the photo on p. 84). The inspector will normally make it a point to verify its presence. Some panels use a screw that is turned down through the grounding bus and into the panel's steel, and others will have a bonding wire that runs from the bus through a prepunched hole in the steel frame. Though the inspector may only check the bond in the main panel, all metal box frames in the electrical system need to be bonded into the grounding system, such as the cutoff boxes for heat pumps, water heaters and submersible pumps.

Attaching Ground Wire with an Exothermic Weld

As a specialist in lightning suppression, I have to make sure my grounding resistance is as close to zero as possible. That's why I prefer to attach ground wire to the ground rods using an exothermic weld, which provides as low a resistance as you can get, and it will last a lifetime. I've taken ground-resistance readings across all types of clamps and have had a high of 140 ohms through the clamp. With an exothermic weld, it is simply 0 ohms, and more importantly, it will stay at zero because there is no corrosion between clamp and rod and no screws to work loose. I use a system called One Shot. Available at some electrical supply houses and costing around $10, this exothermic weld system consists of a mold and a disk that holds the metal weld and igniter powder in the mold. You'll also need a special flint igniter. Here's how it works.

Dig the hole and drive the ground rod. Place the mold onto the rod and insert the disk into the mold. Run the ground wire all the way through the mold if the wire is running to another

Insert the grounding wire into the mold. (Photos courtesy of One Shot)

Pour the shot and igniter powder into mold. Don't mix metal weld and igniter powder; leave them separate.

Grounding Wire Connection

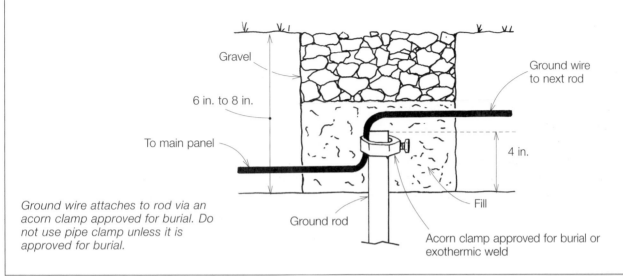

Gravel

6 in. to 8 in.

To main panel

Ground wire
to next rod

4 in.

Fill

Ground rod

Acorn clamp approved for burial or
exothermic weld

*Ground wire attaches to rod via an
acorn clamp approved for burial. Do
not use pipe clamp unless it is
approved for burial.*

rod. If you're attaching the wire to the last rod in the system, run the wire through and end just outside the mold. (Molds also are available with only one hole, but I always buy the ones with two located opposite each other.)

Pour the metal weld powder (called a shot) and igniter powder in the mold. The igniter powder will come out on top of the shot at the top of the mold. The igniter powder is a little lighter in color than the metal weld. Don't mix the igniter powder with the shot. Keep it on top where it's poured, and leave it loose—don't pack it.

While wearing eye and hand protection, put the cover on the mold and hold the flint igniter over the top. Spark the igniter to fire up the shot (it may take two or three sparks to get the fire, depending how concentrated the igniter powder is under the cover hole). Once ignited, stand back and watch the volcano. Don't touch the wire or the rod until it cools, after which you can break off the mold (I leave the mold on). It usually takes about an hour to cool.

Use the flint igniter to flame the weld (wear eye and hand protection).

When rod and wire cool, break off the mold. If you prefer, you can leave the mold on, as the author does.

The metal strap in the bottom right of the photo is the panel bond. The wire above it goes to the ground rods.

Testing

A plug-in circuit analyzer can verify the presence of an equipment grounding wire from a three-prong receptacle to the main panel. But it cannot verify the presence or even condition of the grounding wire going to the ground-rod system, and most won't tell you if a metal box or an appliance has no grounds (for more on plug-in testers, see pp. 43-44).

WHAT TO GROUND

When it comes to grounding, remember this rule of thumb: If it can become accidentally energized—through induction or fault—ground it. I have even been shocked by sheet-metal ductwork both under and in a house. Therefore, I require ductwork to be grounded. Just because the frame of the heat pump or furnace is grounded by its own equipment

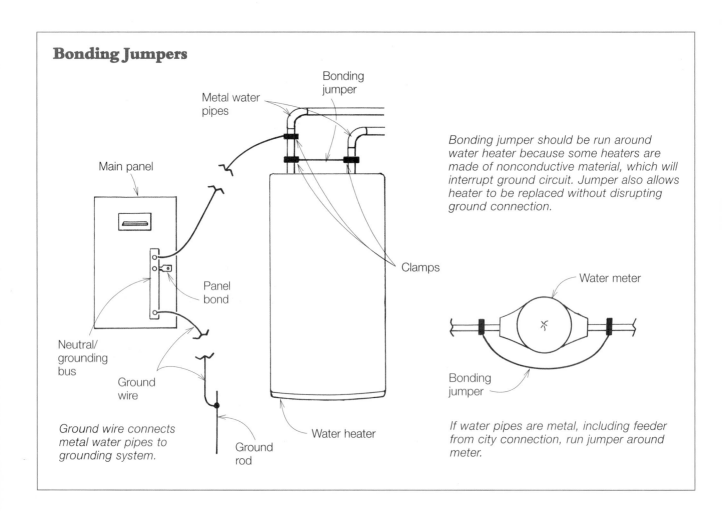

Bonding Jumpers

Bonding jumper

Metal water pipes

Main panel

Bonding jumper should be run around water heater because some heaters are made of nonconductive material, which will interrupt ground circuit. Jumper also allows heater to be replaced without disrupting ground connection.

Clamps

Panel bond

Neutral/ grounding bus

Ground wire

Water meter

Bonding jumper

Ground wire connects metal water pipes to grounding system.

Water heater

Ground rod

If water pipes are metal, including feeder from city connection, run jumper around meter.

grounding wire, doesn't mean that the ductwork is grounded. Many times there is a nonconductive flex joint between the unit and the ductwork that acts as an insulator. In addition, the rails of garage-door openers should be grounded to prevent lightning surges from building up and destroying the electronic opener. This should also apply to any structural steel within the building. Here's a list of things that I recommend be grounded.

Metal Water Pipes

Do not use metal water piping in lieu of ground rods or even as the sole grounding wire, even if local codes allow it. Instead of thinking of a metal water pipe as something to ground to, think of it as something that needs to be grounded. Codes normally allow a grounding connection to a metal pipe if the pipe is in the ground for 10 ft. or more and the connection is made within 5 ft. of where the pipe enters the building. But this grounding method requires a supplemental electrode (a ground-rod system). If you need a ground in a house that has old two-conductor wiring, don't clamp onto the metal water piping, even if local code allows it. It is simply a matter of time before a plumber will replace a section of pipe with plastic—disconnecting any ground connection. So don't use metal water pipes as grounding conductors. When looking for a ground, it's best to run an equipment grounding wire back to the main panel, ground rod or grounding electrode conductor.

To ground metal pipe, make the primary connection from the metal pipe to the grounding bus within the main panel (use a full-size grounding conductor, as shown in the chart on p. 81). You'll also need to run bonding jumpers where needed. The bonding jumper is a separate section of wire with clamps on each end that allows the grounding conductor to be run around anything that could interrupt the grounding circuit (see the drawing on the facing page). For example, you need to run a jumper around the water heater because some water heaters are now made of nonconductive material. Also, water heaters get replaced, and you don't want to disconnect the ground at any time. If all pipes are metal, including the feeder in from the city connection, a jumper will need to be made around the meter.

Metal Conduit

All metal conduit run within the house must be grounded. I normally don't depend on conduit to carry the ground from location to location on any type of slip-in fitting because there is too much of a chance of a loose connection opening the grounding system (screw-in fittings make better grounds). I normally run a grounding wire within the conduit as much as practical. Consider using plastic conduit if you don't want to worry about grounding it.

In a house, metal conduit is most commonly used in the garage and basement. Short sections of thin-wall conduit protect the wire as it runs from the receptacle and switch boxes vertically up to the floor-joist system. To ensure a proper ground, connect the metal conduit tightly to the metal receptacle box and ground the metal receptacle box via the

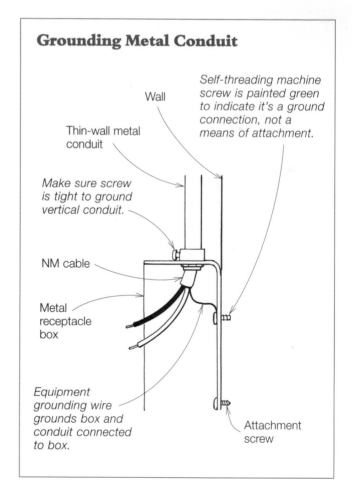

Grounding Metal Conduit

Self-threading machine screw is painted green to indicate it's a ground connection, not a means of attachment.

Wall

Thin-wall metal conduit

Make sure screw is tight to ground vertical conduit.

NM cable

Metal receptacle box

Equipment grounding wire grounds box and conduit connected to box.

Attachment screw

grounding wire in the NM cable entering the box. Use a separate screw to hold the grounding wire against the box—not one holding the box to the wall—and paint the screw green to set it apart from the rest (see the drawing on p. 85). In addition, code forbids the use of sheet-metal screws to hold down ground wires.

Receptacles

Receptacles in nonmetallic boxes can be grounded in several ways (see the drawings below). The standard way is by simply connecting the incoming bare grounding wire onto the grounding screw of the receptacle (drawing A). This, however, only works if there is but one incoming cable. If there are two or more cables, the rules change a bit because only one wire is allowed to be connected under a screw head. Cut both ground wires from the cables the same length, and add a 6-in. bare wire to the bunch, which will serve as a jumper from the splice to the receptacle. Twist the ends of the wires together in a clockwise direction with broad-nosed electrician's pliers for about 1 in. to 1½ in. Then either twist on a wire nut (red or yellow, not green) or crimp on a copper sleeve made for splicing (for more on splicing and wire nuts, see pp. 152-154).

Grounding Receptacles in Nonmetallic Boxes

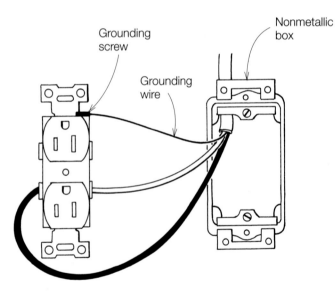

A *Simplest method runs ground from cable directly to grounding screw on receptacle.*

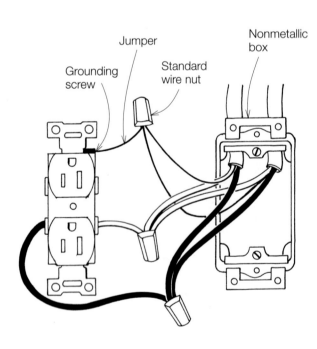

B *When two or more cables enter box, use standard wire nut to splice grounds and jumper, then attach jumper to grounding screw on receptacle.*

Connect the jumper to the receptacle's grounding screw (drawings B and C).

Another method utilizes a green grounding wire nut that has a hole in its end. Cut one grounding wire 6 in. longer than the other and place them parallel with each other, overlapping by 3 in. Then twist them together using broad-nose pliers. About 6 in. from the end of the longer ground, twist the wires together. Slip the green wire nut over the long wire and push it down to the splice and twist it on. Then connect the long wire to the receptacle ground screw (drawing D). In many areas, it is not acceptable simply to twist the wires in the splice; the wires tend to come apart over time without a mechanical connection.

To ground a metal receptacle box, add another jumper to the splice, which attaches to a green grounding screw on the box. These jumpers are called pigtails and can be purchased with screws already attached. Some receptacles have an automatic grounding feature, consisting of a metal clip that keeps the receptacle screw tight against its yoke so as to ground the box by bonding onto a grounded receptacle. This method

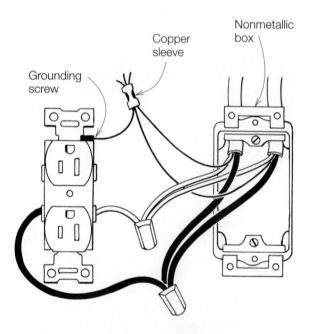

C When two or more cables enter box, splice grounds and jumper with crimped copper sleeve, then attach jumper to grounding screw on receptacle.

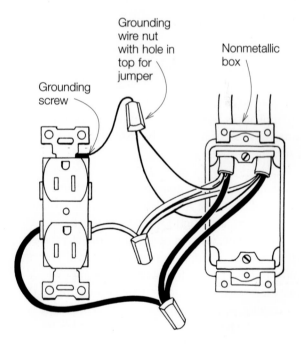

D When two or more cables enter box, use grounding wire nut (green) to splice grounds and jumper, then attach jumper to grounding screw on receptacle.

Automatic Ground

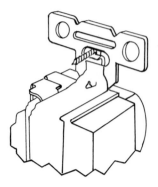

Spring-type grounding strap on high-quality receptacle automatically grounds metal box when receptacle is screwed into it.

Adapter Does Not Ground

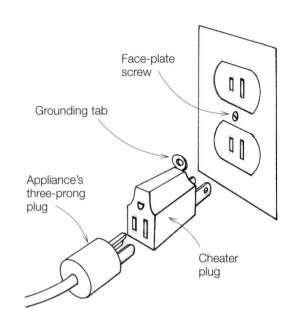

Face-plate screw

Grounding tab

Appliance's three-prong plug

Cheater plug

Cheater plug is only an adapter and will not provide ground simply by plugging it in. It's better to install grounded receptacle and forego adapter.

will eliminate the pigtail jumper that connects the receptacle ground screw to the twisted bunch (see the drawing above).

Appliances

Until the 1996 NEC, some appliances, in particular electric dryers and stoves, were allowed to use the neutral as the ground as well. Now, however, all appliances must use an equipment grounding wire that's separate from the insulated current-carrying neutral.

Most appliances are grounded via their three-prong plug and cord (the roundish prong is the ground connection), although some appliances have their ground connections hard-wired by the electrician. However, for an appliance to be grounded via the plug and cord, a grounded receptacle *must* be installed, with a grounding wire connecting to the grounding system. To make sure the receptacle is grounded, check it with a plug-in tester before plugging in a grounded appliance.

If you're thinking of plugging in a grounded appliance to an ungrounded receptacle using a

cheater plug, don't do it. The cheater plug is simply an adapter that allows you to insert a three-prong plug into a standard outlet (see the drawing above). It will not ground the appliance. If the appliance has a grounded plug, the manufacturer wants the unit to be really grounded—not just plugged into an adapter. And if the appliance does not have a proper ground connection and a ground fault occurs, someone could be electrocuted. A cheater plug will ground the appliance only if the receptacle box is metal, and there is some type of conduit providing a ground all the way back to the main panel. This is rare in residential situations but does occur in older houses wired entirely with BX or Greenfield cable.

Submersible Pumps

If you've got a well, its submersible pump and metal casings are required to be grounded. The metal well casing, around 6 in. in diameter and driven through

Grounding a Submersible Pump

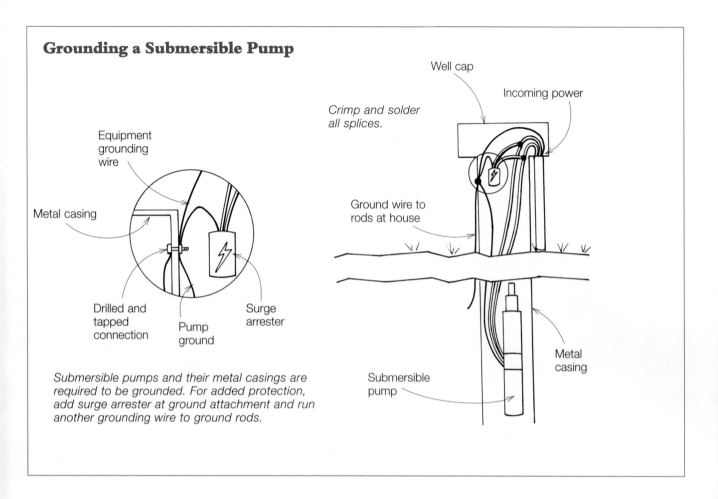

Crimp and solder all splices.

Well cap

Incoming power

Equipment grounding wire

Metal casing

Ground wire to rods at house

Drilled and tapped connection

Pump ground

Surge arrester

Metal casing

Submersible pump

Submersible pumps and their metal casings are required to be grounded. For added protection, add surge arrester at ground attachment and run another grounding wire to ground rods.

the soft soil and into bedrock, will arguably be the best ground medium available, that is, having the lowest resistance to ground of anything around. The NEC requires that you simply attach the equipment grounding wire going to the pump to the metal casing. In high lightning areas, it's best to supplement the code-mandated ground connection by running a 4- or 6-gauge copper wire from the casing to the ground-rod system. I also install a Tytewadd surge arrester at the ground attachment to give surge protection for the pump (see the drawing above).

Structural Steel

Exposed structural steel such as a building frame should be bonded to the grounding system. The bonding jumper, where connected to the frame, needs to be accessible. Though it can connect to the ground rods or to a grounding electrode conductor, I prefer to connect it directly to the main-panel grounding system so I don't have to worry about any bolted connections coming loose. The conductor is sized according to Table 250-94 of the NEC (for 200-amp service, use 4-gauge copper).

Gas Pipes

Whether gas pipes need to be grounded depends on the local code. If the pipe enters the house and is connected to electrical equipment, codes normally require it to be bonded into the grounding system. You can use the equipment grounding conductor of the cable feeding the appliance. If you ground the furnace, for example, and the pipe is screwed into the furnace, it automatically becomes grounded, and you don't have to run a separate ground wire. The NEC forbids the use of a gas pipe as a grounding electrode.

Chapter 6

THE WIRING SYSTEM

LAYING OUT
RECEPTACLES AND
SWITCHES

ROUTING WIRES IN NEW
CONSTRUCTION

ROUTING WIRES IN
RENOVATION WORK

Getting the wires from one point to another in a house is perhaps the hardest part of installing an electrical system. Through trial and error, electricians develop tricks over the years. In this chapter I will show you what I've learned from my experience to make running wire go smoothly. The first and most important step in running wire is planning each branch circuit to make routing the wire easier. (All wires are assumed to be copper unless noted as aluminum.)

LAYING OUT RECEPTACLES AND SWITCHES

The most efficient way to install an electrical system is to lay out all receptacle and switch locations before running the wires. Once these locations are planned, and the cables are run from the main panel to the box locations, it's time to install the boxes.

I normally mount switch boxes 48½ in. from the ceiling, using the top of the double plate as reference. This allows the first piece of drywall (in most houses, finished walls are covered with ½-in. drywall) to go on without any cutting, and the second piece only has to be notched. In a room with a ceiling height greater than 8 ft., measure 48 in. from the floor. I measure, mark with a black felt-tipped pen, then install the box. Unless directed otherwise, I install receptacle boxes one hammer height tall along the floor (see the top drawing on the facing page). This is faster than measuring and just as accurate. I place the hammer on top of the bottom plate, and the box on the hammer. Then I withdraw the hammer and nail on the box. The only thing that differs from room to room is where along the walls the boxes are to be mounted.

Kitchen

There are perhaps more receptacles in the kitchen than any other place in the house (see the drawing below). In general, place receptacles where you think they'll be needed, but remember the basic rules: Outlets are required on any wall space of 2 ft. or more, with one placed within 6 ft. of a door and spaced an absolute minimum of every 12 ft. after that. Good places to install receptacles are on each side of a window (I prefer not to install receptacles under a window) and on each side of a corner. Never install a receptacle above a baseboard heater.

Shortcut to Find Box Height

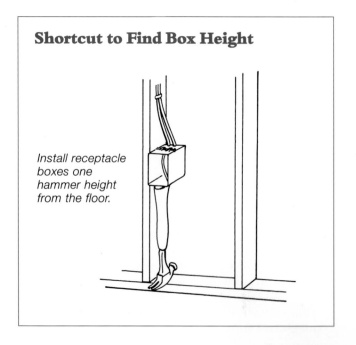

Install receptacle boxes one hammer height from the floor.

Kitchen Receptacles and Switches

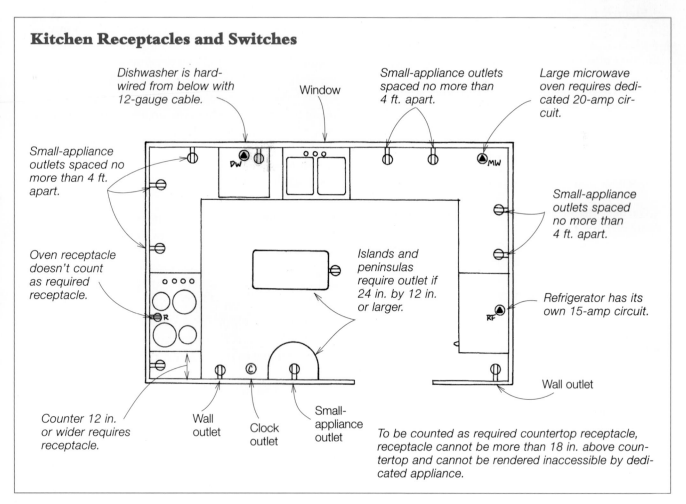

Dishwasher is hard-wired from below with 12-gauge cable.

Window

Small-appliance outlets spaced no more than 4 ft. apart.

Large microwave oven requires dedicated 20-amp circuit.

Small-appliance outlets spaced no more than 4 ft. apart.

Small-appliance outlets spaced no more than 4 ft. apart.

Oven receptacle doesn't count as required receptacle.

Islands and peninsulas require outlet if 24 in. by 12 in. or larger.

Refrigerator has its own 15-amp circuit.

Counter 12 in. or wider requires receptacle.

Wall outlet

Clock outlet

Small-appliance outlet

Wall outlet

To be counted as required countertop receptacle, receptacle cannot be more than 18 in. above countertop and cannot be rendered inaccessible by dedicated appliance.

Gauging Branch Circuits

Each area of the house will incorporate wiring of different gauges, with 14 gauge being the smallest. Unless you are in a tight, low-bid situation, don't use 14 gauge—even though it's cheaper—because it is easy to overload. I always use 12-gauge wiring for general-purpose circuits and all lighting circuits, and heavier gauge for special applications, such as an electric stove, dryer or baseboard heater. Using 12 gauge allows for a little extra load on the circuits and also allows for upgrading. The following generalized list will help you choose the correct wire gauge for a particular circuit in a typical residential installation.

• **Bathroom receptacles: 12 gauge only (home run—wires from main panel)**

• **Electric baseboard heat: 12 or 10 gauge**

• **Electric dryer: 10 gauge (three insulated conductors plus ground)**

• **Electric stove: 6 gauge (three insulated conductors plus ground)**

• **Electric water heater (4,500 watt, 240 volt): 10 gauge**

• **Garbage disposal: 14 or 12 gauge**

• **Garbage disposal and hot-water dispenser on same circuit: 12 gauge**

• **General-purpose receptacles: 14 or 12 gauge**

• **General-purpose lighting: 14 or 12 gauge**

• **Hot-water dispenser: 14 or 12 gauge**

• **Laundry area: 12 gauge (home run)**

• **Shop circuits: 12 gauge (home run)**

• **Kitchen/dining/pantry: 12 gauge (two minimum, home run).**

Any counter 12 in. or longer requires a receptacle. The countertop requires a minimum of two 20-amp circuits, which may be shared with other receptacles in the dining room and pantry. A kitchen peninsula or island larger than 24 in. by 12 in. requires a receptacle. All kitchen-counter receptacles must be GFCI-protected, including peninsulas and islands (I'll talk more about ground-fault circuit interrupters in Chapter 8). An outlet must be within 2 ft. of the counter edge, and no farther than 4 ft. apart after that. The receptacle cannot face up. A good design for the kitchen is to place a box at every other stud along the countertop, which puts a receptacle every 32 in. If more outlets will be required, place a box on every stud. Receptacles located in appliances and fixtures do not count as required receptacles. Neither do hidden receptacles inside cabinets and cupboards, nor those located over 5½ ft. above the floor.

The standard countertop height is 40 in. to the top of the backsplash. I place receptacles 2 in. to 3 in. above the backsplash. The distance the box is to be mounted out from stud will depend on the finish for the wall, whether it will be tile or drywall. If the finish has not been determined by the time you're ready to wire the kitchen, use an adjustable box. This type of box will allow you to move the box out to the finished wall with a simple screw adjustment after the box is nailed on (for more on this box, see p. 92).

Place a receptacle box low for the refrigerator. The 1996 code allows the refrigerator to be on a 15-amp circuit, a change from previous codes. If needed, place a receptacle for the garbage disposal under the sink. A dishwasher normally requires its own circuit and usually just comes up from the floor or right out of the drywall behind the dishwasher. A large microwave oven is normally required—by the manufacturer—to have its own circuit as well.

Place switched lighting at each entrance to the kitchen, and an optional overhead light at the kitchen sink or under the counters. It's also a good practice to place a multigang box or power strip in the counter area, where a cluster of small appliances will be located.

Dining Room

The dining room needs switched lighting at each entrance. The standard rule of placing receptacles applies here. All receptacles within the dining room, the kitchen and the pantry are to be within themselves. That is, the circuits are not allowed to be shared by circuits in other rooms. You are no longer allowed to attach an outside outlet onto the kitchen, dining or pantry circuits. Even though a bedroom shares the dining-room wall, it cannot tap off one of the dining-room's receptacles. Sliding glass doors are special. Treat the slider as a door, and the fixed glass as a wall. It's a good idea to place receptacles on each side of the windows for electric decorations that may be used during holidays.

Bathroom

The bathroom now requires its own 20-amp GFCI-protected receptacle circuit. Place a receptacle within 12 in. of a single sink. Double sinks are treated differently: Place one receptacle between the bowls, or one on each side of the sink. If the bath doesn't have a window, a circuit for a vent fan will need to be installed. If the fan is in the shower/bath area, it must be listed for wet locations and must be GFCI-protected (see the drawing at right).

The bathroom needs switched lighting, which cannot be put on the receptacle circuit. It's also a good idea to have switched lighting at the sink to reduce shadows. Any receptacle outlets in lights or medicine cabinets of the old, two-prong, ungrounded variety should be cut dead. Any of the three-prong grounded variety must be cut into the GFCI system or be cut dead as well.

Fan/light/heat combos need an insulated conductor for each independent function. A fan/light combo with the fan and light operating independently needs a standard three-conductor cable, such as 12-3 with ground. The red and black wires feed the fan and light, the white is the common neutral and the bare ground for the frame. Unless you can come up with four-conductor cable, a fan/light/heat combo will need both a three-conductor cable for the fan and light, and an additional 12-2 cable with ground for the heater. No hanging fixture or lighting track

Bathroom Receptacles and Switches

Vent fan listed for wet locations and GFCI-protected

GFCI-protected outlet

GFCI-protected outlet

Sink light switch

Fan switch

Overhead light switch

Bath receptacles are powered off dedicated 12-gauge circuit. Lights should not be on bath circuit receptacles.

can be located over the tub, unless it is more than 8 ft. above the tub, nor can it be located within 3 ft. of the outside of the tub.

Hallways and Stairs

One receptacle per hallway of 10 ft. or more is required, but none is required for the stairs. However, it is a good idea to install one receptacle at the top, in the middle and at the bottom of the stairs. A hallway and stairs with six steps or more will require three-way switched lighting at each end (for more on three-way switches, see pp. 174 and 175).

Bedrooms

Bedrooms require switched lighting via either an overhead light or by a switched receptacle. A good design will go beyond the minimum number of receptacles. Place receptacles on both sides of windows, and add receptacles for table lamps and any electronic gear that might be located in the room. A three-way switch next to the bed will allow a person to turn off the overhead light without getting out of bed. Table lamps, or even track lighting, on each side of the bed will give an independent light source for reading.

Closet Lights

Closet lights require some design thought before installation. No longer are you allowed simply to wire in an open bulb with a pull chain or any type of hanging light. In fact, exposed incandescent bulbs or even partially exposed are not allowed. There is too much of a chance of the bulb breaking and falling on flammable material. Both exposed and enclosed fluorescent fixtures are allowed.

All lights can be fully enclosed, recessed or surface-mounted. The standard location for mounting a closet light is right above the door, facing into the closet. Using this location as a reference, surface-mounted incandescent fixtures must be 12 in. from the storage area (measured horizontally); surface-mounted fluorescent fixtures require 6 in. Recessed incandescent and fluorescent fixtures must be 6 in. from storage (see the drawing on the facing page).

Living Room

There isn't anything special about wiring a living room. It needs switched lighting, preferably at each entrance. The lighting can be overhead or a switched table lamp. Place receptacles within 6 ft. of a door and the minimum of every 12 ft. after that. It's a good idea to place receptacles on both sides of a window and on both sides of a corner and fireplace. And be sure to place extra receptacles for the stereo, VCR, television, satellite system and any other electronic gear that might be used in this room. It would also be a good idea to put the electronic gear on its own 20-amp circuit.

Garages and Unfinished Basements

An attached garage and an unfinished basement are required to have at least one GFCI-protected receptacle and light (an unattached garage does not require these, unless power has been brought in). Garages and basements may have workbenches, so be sure to install a few receptacles at workbench height (around 48 in.). Any accessible general-purpose receptacle must be GFCI-protected, unless the receptacle is for a dedicated purpose, like a freezer, or it's inaccessible, like that for a garage-door opener. The dedicated receptacle must be a single receptacle or have both inputs exempt from the GFCI requirement.

If the garage shares a wall with a room in the house, never install back-to-back receptacles within the same wall cavity. In a garage, the drywall is a fire barrier. If you cut a hole in the fire barrier in the garage and then cut another hole inside the house in the same wall cavity, you open up a fire entry into the house. It's especially important to avoid installing a panel in this wall, too. This is a common and expensive mistake.

Laundry Room

A 20-amp circuit is required for the laundry-room receptacles and cannot be a shared circuit with any other room. You can get by with just one receptacle if that's all you want. Install switched lighting off a different circuit at each entrance. (The dryer is an independent appliance and is not considered part of the laundry circuit.)

Attics, Crawlspaces and Outside

If the area contains maintainable equipment, such as a water heater or a heat pump, there must be a switched light. Any crawlspace receptacle must be GFCI-protected.

Every house has to have at least two outside receptacle outlets: one for the front of the house, and one for the back—many homes have more. All outside receptacles must be GFCI-protected.

Closet Lighting Minimum Clearances

Code-approved installations

Surface-mounted incandescent fixture with enclosed lamp(s)

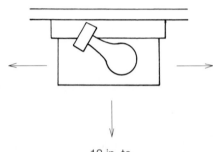

12 in. to
stored material

Surface-mounted exposed fluorescent fixture

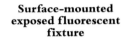

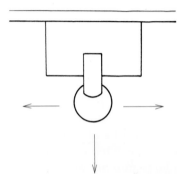

6 in. to
stored material

Recessed incandescent fixture with enclosed lamp(s)

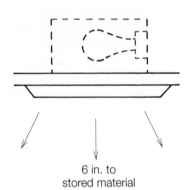

6 in. to
stored material

Recessed fluorescent fixture

6 in. to
stored material

Closets do not have to have lights. But any lights must be kept away from storage materials. Exact minimum distance will depend on type of light. If lights are mounted on ceiling, distance reference is to edge of fixture, not bulb. However, if bulb protrudes beyond fixture base, it is safer to use bulb as reference point.

Violations

Incandescent lamps all or partially exposed	Hanging fixtures	Hanging lamp holders

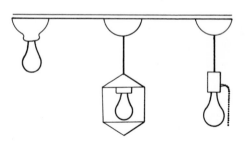

ROUTING WIRES IN NEW CONSTRUCTION

Once the main service panel is mounted, you'll need to open holes through the plates immediately above and below the panel. (Remember to punch out the knockouts in the top and bottom of the panel; I usually do that before mounting the panel.) All cables from the main panel pass through these holes. You've got a few options to get the cables where they need to go: You can route them through the crawlspace or the attic, and through, or over, studs and joists. For neatness and safety, try to keep the main runs along the outer edge of the building.

Maintaining Structural Integrity

No matter where the cable is run, the structural integrity of the wood must be of prime consideration. That is, do not drill a large number of large-diameter holes throughout the framing member to pass cable through. Local codes dictate whether or not the structural integrity of any framing member—stud, plate, joist—is being violated. If you have any doubts, it's a good idea to check with the local inspector or lead carpenter before drilling or cutting. Some common codes for retaining structural integrity are as follows:

Studs

Notches in bearing-wall studs should not exceed 25% of the stud width, and holes should not exceed 40% of the stud width. For a 2x4 stud, the notch would have to be approximately 7/8 in., and the hole diameter cannot be greater than 1½ in. Studs in nonbearing walls can be notched or drilled up to 40% of the stud width. All holes must be drilled exactly in the center of the stud. If you need to drill an additional hole, locate it above or below the first, keeping all drilling to the exact center of the stud.

If the hole is within 1¼ in. of the edge of the stud, a steel plate is required to be installed. The plate protects the wiring from screws and nails from drywall and siding, which will be driven after the wiring is roughed in.

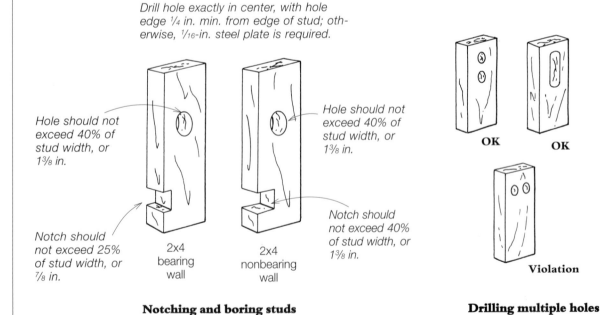

Drill hole exactly in center, with hole edge ¼ in. min. from edge of stud; otherwise, 1/16-in. steel plate is required.

Hole should not exceed 40% of stud width, or 1⅜ in.

Hole should not exceed 40% of stud width, or 1⅜ in.

Notch should not exceed 25% of stud width, or 7/8 in.

2x4 bearing wall

2x4 nonbearing wall

Notch should not exceed 40% of stud width, or 1⅜ in.

OK   OK

Violation

Notching and boring studs

Drilling multiple holes

When I'm sure that structural integrity of the plates won't be compromised, I open up a large slot—as opposed to drilling a series of holes—to make passing the cables from the main panel easier (for more on maintaining structural integrity, see below). To open the slot, drill two ⅞-in. holes about 12 in. apart and cut out the area between with a reciprocating saw

(see the drawing on p. 98). Be sure to wear safety glasses when drilling and cutting. Make the same cut through the bottom plate to run cable into a crawlspace or basement. Maintaining structural integrity is rarely a problem when cutting through the plates in this manner. But if you've got any doubts, ask the lead carpenter on the job site if the

Similarly, all notches in studs must be covered with a steel plate. The plate must be at least 1/16 in. thick and must be flush to the stud edge (homemade plates can be used, as long as they meet the minimum thickness).

Plates and joists

Some building codes say that if a cutout in the top or bottom plate reduces the plate by more than half its original width, a steel plate must be installed to strengthen the plate. However, this is mostly ignored. Notches in joists should not exceed 25% of the joist width. Holes should not be located within 2 in. of the top third or bottom, and the diameter should not exceed one-third of the width of the joist. Never notch or drill decorative timbers, like those in a timber frame, and never notch or drill large beams without consulting the lead carpenter.

Cable

Never staple a cable on its side because the staples will damage the conductors and tear the sheath. Never staple a cable extremely tight against the stud because it could short across the conductors. Instead, staple only to a snug fit. Nonmetallic cable should be supported a minimum of every 4½ ft., but I prefer to staple it every 2 ft. to 3 ft.

When stud is notched or bored, code requires 1/16-in. steel plate to protect cables from screws and nails.

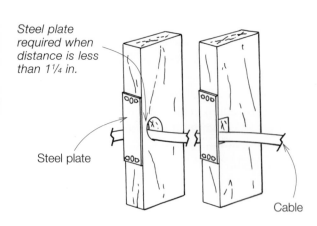

Steel plate required when distance is less than 1 1/4 in.

Steel plate

Cable

Protecting cables

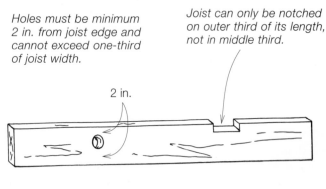

Holes must be minimum 2 in. from joist edge and cannot exceed one-third of joist width.

Joist can only be notched on outer third of its length, not in middle third.

2 in.

Notching and boring joists

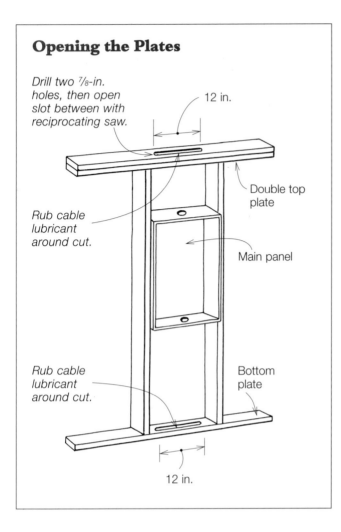

Opening the Plates

Drill two ⅞-in. holes, then open slot between with reciprocating saw.

12 in.

Double top plate

Rub cable lubricant around cut.

Main panel

Rub cable lubricant around cut.

Bottom plate

12 in.

(the NEC gives guidelines for derating conductors in Article 310, Note 8). To avoid these adjustments, simply don't bundle the cables together. It will make your life easier. I once went to a job that had failed inspection, and the owner was paying me to bring everything up to code. It was the neatest job I had ever seen—even the stranded neutral had been braided like a girl's pigtail. The owner ran *all* the wires—10 cables (20 or more individual conductors)—through the same large hole and bundled everything tightly with tie wraps for the entire run. It looked great, but it failed inspection because the cables, especially the ones going to the kitchen and laundry, could not be derated.

Through Basements and Crawlspaces

In a basement that will not be finished, NM cable of 8 gauge and larger can be run across the bottom edges of the floor joists, or you can nail a running board under the joists and staple the cables to the board. Cables smaller than 8 gauge run across the bottom of the joists must be stapled to a 1x running board (see the drawing on the facing page).

For a basement that will be finished (either now or in the future), run the cable through the floor joists or along one of them. When running cable along the joist, staple it around every 3 ft. to the center of the face to avoid nails and screws that will be driven from below when the ceiling is installed. When boring holes to run cable through joists, follow the guidelines shown in the drawing on p. 97.

Here's how to pull the cable from the main panel, through the basement or crawlspace to the outlet.

1. Lay the box of cable in front of the main panel and roll out the approximate amount you'll need. Try to unreel the cable from the outside of the roll so that it will remain flat, as opposed to taking it from the center and having it loop in spirals. Next, feed one end of the cable through the lubricated slot in the bottom plate.

2. Pull the cable through the holes in the joists or across the joists (the method you choose depends on

cuts would compromise the plates' structural integrity. Once cut, smear some cable lubricant—available at electrical supply houses—around the inside of the hole. This will allow the cables to slide through smoothly.

Don't bundle groups of cable together for long runs. It may look neater, but if bundled tightly together for 2 ft. or more, the heat generated within the bundle may damage the wire's insulation. This means that you may have to lower the amount of current flowing through the wires (called derating) to lower the temperature and thus protect them. For a bundle of 10 to 20 conductors (five to ten 12-2 w/g cables), for instance, a 12-gauge cable would have to be fused at 15 amps, and a 10-gauge cable at 20 amps

Running Cable in the Basement

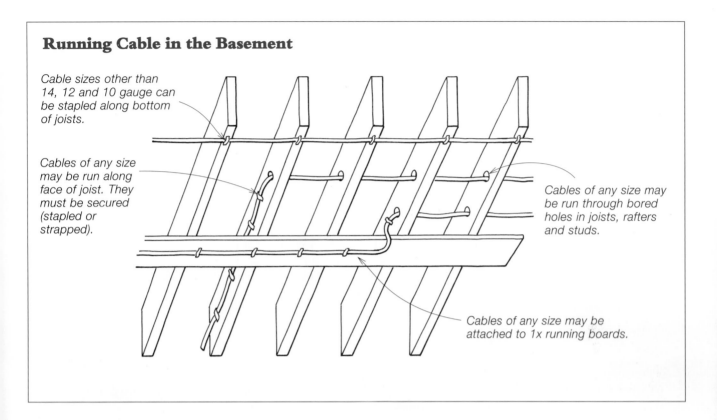

Cable sizes other than 14, 12 and 10 gauge can be stapled along bottom of joists.

Cables of any size may be run along face of joist. They must be secured (stapled or strapped).

Cables of any size may be run through bored holes in joists, rafters and studs.

Cables of any size may be attached to 1x running boards.

Pulling Wire Through Conduit

When pulling several wires through conduit, you need to be very careful because the individual wires can get tangled easily. Professionals use special reel caddies to offset this problem. You are not supposed to pull single sections of the wire at a time, even though it can be done if you're very careful. Instead, as you assemble the conduit, run a pull wire through each section. Once the conduit is assembled, tape all the wires to the pull wire, smear cable lubricant (available at electrical supply stores) all over the wires, and then pull them through the conduit.

Wiring Through Trusses

If trusses are used in the construction, be wary. Trusses can be in the attic or even between floors. Do not drill through or cut trusses in any way (unless approved by the truss manufacturer and building inspector) to run cable because it could ruin their engineered specs and void the manufacturer warranty.

A floor truss has an open area in its design around 12 in. to 18 in. tall. The space is perfect for running wiring across them without the need to drill through the trusses or to install a running board (it's okay to staple wire to the edge of the truss members). Attic trusses, however, have larger open spaces and may be used in accessible areas. If you need to run wire across the trusses, don't drill or cut the trusses; simply install a running board on top near the truss edge and staple the cables on top of the board. (The truss edge is located above the top plate along the outside wall.)

Routing Cable Through an Attic

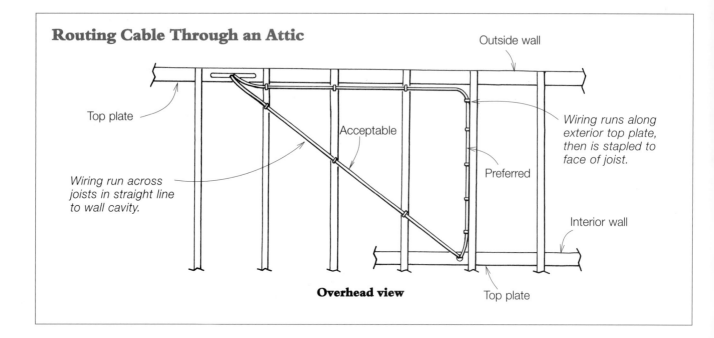

Outside wall

Top plate

Wiring run across joists in straight line to wall cavity.

Acceptable

Preferred

Wiring runs along exterior top plate, then is stapled to face of joist.

Interior wall

Overhead view

Top plate

whether the basement will have a finished ceiling) to the wall cavity in which the cable will terminate.

3. Drill through the bottom plate of the interior wall and then pull enough slack up and to the location of the receptacle or switch.

4. Staple or attach the cable wherever needed in the basement (along the face or edge of the joists or on the running board, every 3 ft., not the code minimum of 4½ ft.).

5. At the main panel, cut the cable with enough slack to go from the large hole in the bottom plate to the top of the main panel.

6. Mark the cable to indicate where it goes.

Through Attics

There are two different routes to run cable through an attic: across the joists in a straight line, or over the joists near the edge (above the plate of the outside wall) and then along the face of one joist to the wall cavity (see the drawing above). Running cables in a straight line across the ceiling joists from main panel to wall cavity is fast and will require less

cable. But the method shouldn't be used in an accessible attic because the cables will get in the way of people walking in the attic, and they will also be in the way if any flooring goes down. In an accessible attic, I run the cable along the outside walls, over the tops of the ceiling joists or truss edges, then along the face of a ceiling joist. Because of low headroom, the outside edges of the joists are normally inaccessible, even in an accessible attic, so running cable over them is not a problem. In attics accessible by permanent stairs or a ladder, wiring needs to be protected with a guard strip that's at least as high as the cables to prevent people from tripping on them. Where there is no permanent stairs or ladder, this protection is needed only for the first 6 ft. around the scuttle-hole entrance.

Once you've got the wiring route through the attic planned out, you can begin to pull the cables. Here's how (this method can be used for joists or trusses).

1. Lay the box of cable in front of the main panel and roll out the approximate amount you will need from the outside of the roll. Work out the coils so that the cable will lie flat. Next, feed one end of the cable up and over the first ceiling joist. Don't insert

the cable through the large hole in the double top plate above the panel yet. Doing so will make it much harder to pull it over the top of the joists and could damage the cable because of the sharp 90° turn it has to make immediately above the top plate.

2. Pull the cable over the joists to whatever stud-wall cavity you need to get into. Wiring home runs (the cables from the main panel) along an exterior wall allows you to keep the wiring out of harms way in the attic. To get cable to an interior partition wall, simply pull enough slack through to make the run along the edge and then along the face of the ceiling joist to the cavity.

3. Once at the correct wall cavity, drill through the top plate and pull enough slack to get to the intended receptacle or switch.

4. Insert the cable through the hole in the top plate and pull the slack all the way through to its intended location. In the attic, staple the cable wherever needed (along the face or edge of the joists, every 3 ft., not the code minimum of 4½ ft.).

5. At the main panel, cut the cable with enough slack to go from the large hole in the double top plate above to the bottom of the main panel.

6. Insert the cable from the attic through the double plate and into the main panel.

7. Mark the cable to indicate where it goes.

Along and Through Studs

When wiring vertically along a stud, keep the cables to the center of the stud (keep 1¼ in. of space on either side of the wire) and staple them every 2 ft. to 3 ft. If several cables are to be run in a very narrow location, most inspectors will allow two cables under one staple. But if you need to stack more cables, or there isn't enough room under the staple, stack the cables through special plastic brackets found at electrical supply houses. If you need to drill through the studs, follow the guidelines on pp. 96-97 to maintain the structural integrity of the wood. In addition, be sure to drill at least

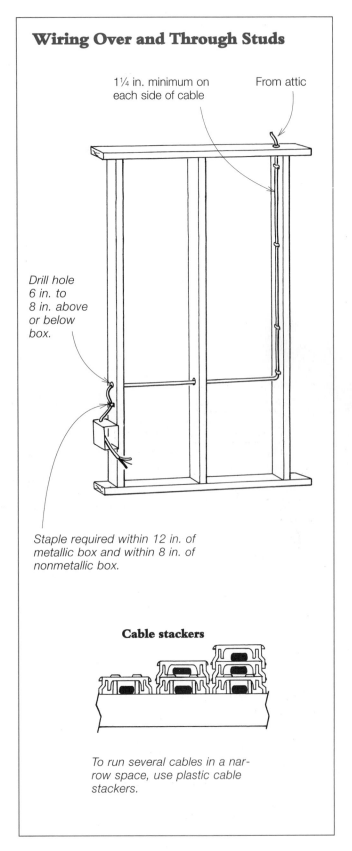

Wiring Over and Through Studs

1¼ in. minimum on each side of cable

From attic

Drill hole 6 in. to 8 in. above or below box.

Staple required within 12 in. of metallic box and within 8 in. of nonmetallic box.

Cable stackers

To run several cables in a narrow space, use plastic cable stackers.

Wiring Around a Window

To get cable around window, run it up through attic, through studs below window, or through basement or crawlspace.

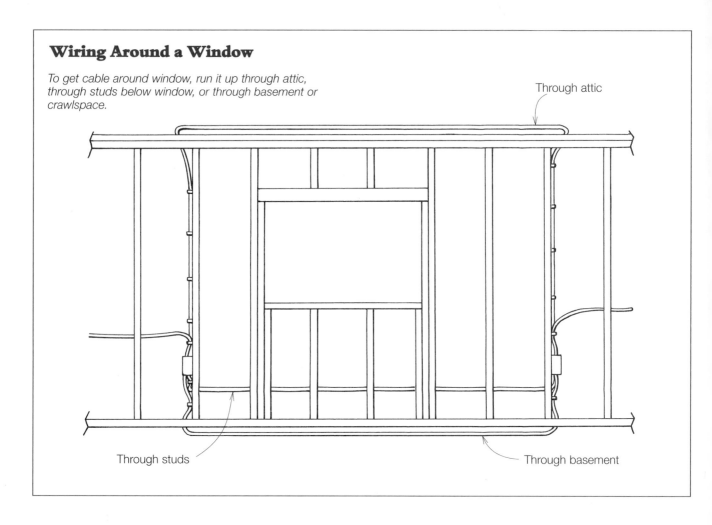

Through attic

Through studs

Through basement

6 in. to 8 in. above or below the receptacle or switch location because code requires the cable to be stapled to the stud within 12 in. of a metal box and within 8 in. of a nonmetallic box (see the drawing on p. 101).

Around Windows

Running cables around windows takes some forethought. Your first thought may be simply to drill every stud that gets into your way, including the triple stud on both sides of the window. But this method requires drilling six to eight times. Normally it's a better idea to take the cable up through the double top plate, through the attic along the exterior wall, and then drop down again through the top plate to the wall cavity on the opposite side of the window. If you prefer, run the wiring through the basement. With the latter two methods, you only need to drill twice (see the drawing above).

Once in a while you may encounter a window that runs from top to bottom with no place to run the cable overhead due to an exposed-beam design. If there is no crawlspace, your first option is through the shim area between the header and the window frame. If this is nonexistent, you'll need to make way for the cable using a router or a circular saw. Set the bit or blade to a depth no deeper than ½ in. (⅜ in. is better) and create a channel across the header for the cable to lie in. Once the cable is in, cover it all the way across with a ¹⁄₁₆-in. steel plate to protect it from nails and screws.

Wiring Around a Door

To get wire around a door, run it through attic, through basement or crawlspace, through cripple studs above header or through shim area of door.

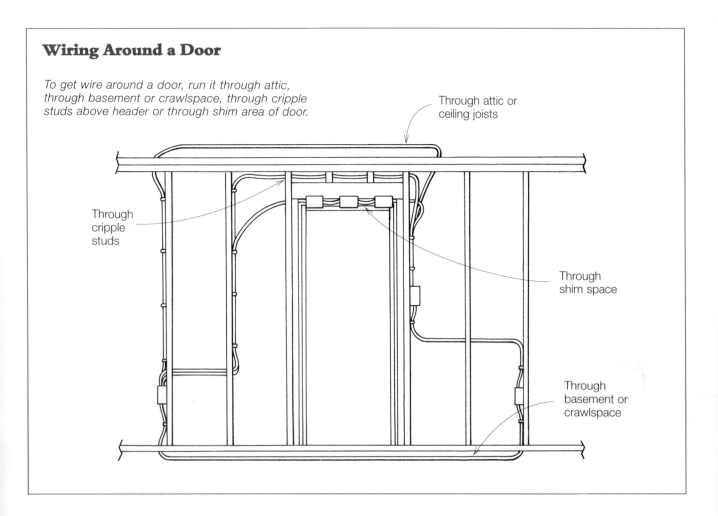

Through attic or ceiling joists

Through cripple studs

Through shim space

Through basement or crawlspace

Around Doors

Getting cable around doors also takes some forethought, but the logic is the same as getting around windows. The easiest route is to go around by going up the stud and through the attic or ceiling joists, or you can go through the basement or crawlspace (see the drawing above). In these situations, you'll only have to drill twice. Another option is to run the cable through the cripple studs above the header, but this method requires a lot of drilling, and sometimes the door header is so high that there's not enough room over the door. Yet another option is to run the cable through the shim area between the door frame and the header and protect the cable with a ¹⁄₁₆-in. steel plate.

Once in a long time, you'll encounter a house with an exposed-beam design and doors with solid headers with no shim room. Whenever there is absolutely no place to run a cable, use a router or a circular saw to create a channel in the header. And don't forget the ¹⁄₁₆-in. steel plate.

Boxed Corners

Boxed corners are always problematic. If the boxed corner looks like a major problem, try to run the cable over and around it (see the left drawing on p. 104). Don't waste your time drilling and fishing the cable through it. If you cannot go around, which is rare, note if the box is hollow or solid.

Hollow boxed corners A hollow boxed corner is easy to get through. Using a 1-in. bit, drill from both

Going Over Corners

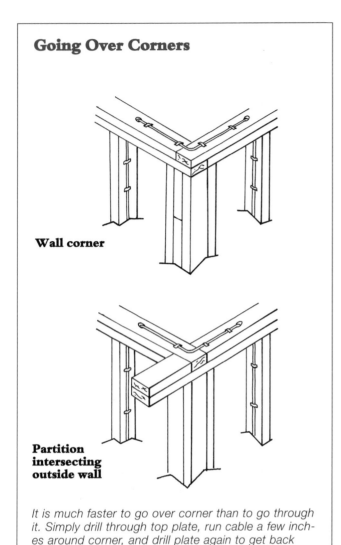

Wall corner

Partition intersecting outside wall

It is much faster to go over corner than to go through it. Simply drill through top plate, run cable a few inches around corner, and drill plate again to get back into wall cavity.

Wiring Through Hollow Corners

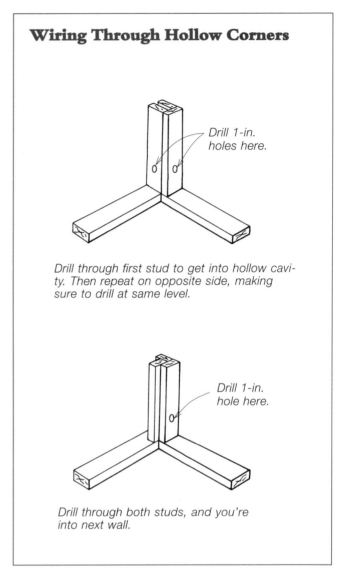

Drill 1-in. holes here.

Drill through first stud to get into hollow cavity. Then repeat on opposite side, making sure to drill at same level.

Drill 1-in. hole here.

Drill through both studs, and you're into next wall.

sides at the same level (see the right drawing, above). Then it's a simple matter of fishing the cable through the holes. If a stud is in the way, and you can't get the drill into the wall cavity (and if the carpenters don't mind), remove it temporarily. Carefully cut the nails at the top and bottom of the stud with the reciprocating saw, and lift the stud out. Cut parallel to the top and bottom plate, and don't ruin the stud by bashing it out with a hammer—the carpenters may massacre you. When you're finished, put the stud back in place.

Solid boxed corners To get through a solid boxed corner, first remove any studs that are blocking your access with the drill. Next, drill a 1-in. hole a little over half way into the corner. Be careful while drilling because the drill is working hard and could buck or stall, and there are a lot of nails in a solid corner. A ⅜-in. drill or even a battery-operated one may be used, but it will take longer and put a lot of wear and tear on the tool. Do not use a pistol-grip ½-in. drill because if it stalls in the wood or hits a nail, the drill could break your wrist. If you use a spade bit for this operation, be sure it is very sharp.

Fishing Cable Through a Solid Corner

Fish wire

1-in. hole

Fish wire

Hook wire

Fish wire

Hook wire

Cable taped to fish wire

1. *Insert fish wire into hole.*

2. *Hook fish wire with another single wire and pull it through.*

3. *Lubricate hole, and pull hook and fish wire and cable through hole.*

Once the hole on one side is complete, drill from the other side so that the holes will meet.

Once the drilling is done, fish the cable through, as shown in the drawing above. This is easier said than done. First, be sure to clean all the wood shavings out of the hole before you start, and put some cable lubricant into the hole to make the cables slide easier. Insert a single wire through one hole to the very end. The single wire will serve as a fish wire. Then, using a stiff wire with its end formed into a hook, reach from the hole on the other side and hook the fish wire. At the same time, pull and feed the fish wire until it comes out the other side of the corner. If it falls off the hook, reach in with some needle-nose pliers, grab it and pull it the rest of the way out.

Drill a hole (at the same height as the holes in the boxed corner) through each stud you had to remove earlier and nail it back in place. Pull the cable through the boxed corner and all appropriate studs. Be careful not to damage the cable as it is pulled through the 90° bend. It would be wise to insert a second fish wire through the hole as the cable is being pulled through, just in case it is decided later

in the job that a second cable needs to get around that corner.

Concrete Walls and Floors

Getting cables through concrete without the proper tools is very difficult. I use a rotary hammer with expensive carbide-tipped drill bits to drive through solid concrete. However, a simple star drill, or even just a hammer, will punch a hole through hollow concrete block.

There are three methods of running cables through basement walls of cinder block or solid concrete. The least desirable way is running them through the inside of the walls, with the boxes installed in cavities broken into the blocks. This, of course, cannot even be considered with poured walls. The most common method is simply to nail and glue pressure-treated boards where the conduit and boxes are to be installed. Then attach the conduit, cables and boxes to the wood. Otherwise, you'll wind up drilling and inserting some type of anchor for every box and anchor point for the conduit—and that's a lot of drilling and anchors to install. If the basement is to have framed walls flush with the concrete walls, the wiring can be run through the framed walls.

Never pour concrete around a wire or cable; always install conduit for the wiring. If outlets need to be in the middle of a concrete floor, which may be the case in a garage or shop, run the conductors through conduit before the pour, with the conduit sticking out above the floor. Avoid installing boxes flush with the floor because water could pour into the outlet, or trash could collect there.

ROUTING WIRES IN RENOVATION WORK

Renovation is not easy. Rewiring a house is very costly. It's very labor-intensive, sometimes taking half a day just to replace or add a single receptacle or switch. I try to make it a point to run the wires in such a way that I make as few holes in the walls as possible. The easiest and cleanest routes are through the attic and basement, but these routes are not always available.

One of the first things you'll have to learn is how to find studs behind a finished wall, which is an art within itself. I've never had any luck with a "stud-sensor" tool, so I can't recommend it. I normally knock on the wall with my knuckles or a screwdriver handle covered with a clean cloth. This, however, will not work for plaster on lath walls. Or you can drill a $\frac{1}{32}$-in. or $\frac{1}{16}$-in. hole or drive in a finish nail until you feel solid wood as opposed to hollow cavity. Another option is to drill a $\frac{1}{4}$-in. to $\frac{3}{8}$-in. hole in the wall, insert a wire bent to a 90° angle, then rotate the wire until you bump the stud. If the trim can be removed, cut into the wall behind the trim to locate the studs. Another trick is to remove a switch or an outlet cover and see which side of the box is nailed to the stud. Then go every 16 in.

Opening a Finished Wall

There are several ways to open a finished wall. You may be tempted to use a jigsaw or reciprocating saw to open a wall. But I do not recommend this because of the high probability of cutting the existing wiring or even the plumbing in the wall. If you know for a fact that there are no wires or plumbing within a particular wall, then use a power tool to open the wall. But make sure the tool is

equipped with a new blade with many small teeth to reduce the pulling, cracking and splintering of the finished wall. To keep the saw's bottom plate from scuffing the wall, cover the plate with tape. Try to cut through just the wallboard. You can adjust the depth of the saw cut by putting a spacer block under the saw's plate, which will extend the blade into the wall cavity just enough to make the cut through

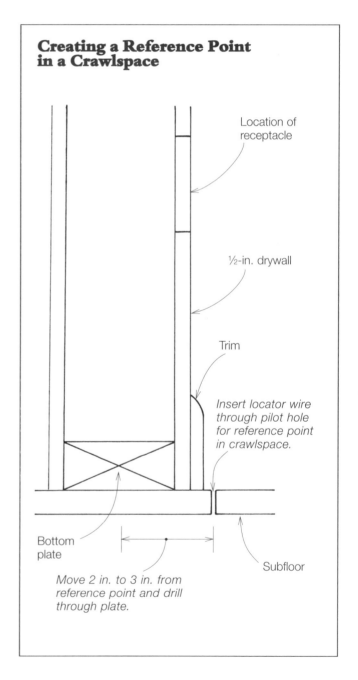

Creating a Reference Point in a Crawlspace

Location of receptacle

½-in. drywall

Trim

Insert locator wire through pilot hole for reference point in crawlspace.

Bottom plate

Move 2 in. to 3 in. from reference point and drill through plate.

Subfloor

the wallboard. However, the preferred method is to use a sharp utility knife or a keyhole saw. A slow, manual cut is much safer and neater than a fast cut with a power tool.

Plaster on lath presents special problems. Many times it splits and produces horizontal cracks in the wall no matter how careful you are. Old molding has the same problems. If the molding is unique, be aware that if it splits, you may not be able to get a replacement piece like it.

When drilling into walls, any drill will do, but it's best to use a cordless drill than an AC-powered one. If the drill bit of a cordless drill cuts into a hot wire, the drill, being mostly plastic, won't be able to electrocute you. Also, never use an old, ungrounded metal-handled drill, unless you have a death wish. And if you use a corded drill, make sure it is powered from a GFCI circuit or has a GFCI-protected extension cord attached.

Wiring Through the Basement or Crawlspace

To run new cables through the crawlspace or basement, you first must locate where they will run to the new receptacle or switch. The simplest way to do that is by first drilling a $\frac{3}{16}$-in. pilot hole in the floor in front of the receptacle or switch in the wall (see the drawing on the facing page). If there is carpet on the floor, use a utility knife to cut a small X where you'll be drilling to keep the carpet strands from getting caught in the drill bit. Once you've drilled through, either leave the bit in the hole or pull the bit out and insert a piece of wire through the hole to the crawlspace, which will give you a reference point in the crawlspace. Move about 2 in. to 3 in. from the reference point, find the bottom plate and drill through the subfloor and plate into the wall cavity. To help you find the bottom plate, look for nails extending through the floor that attach the plate to the subfloor. If you're replacing or rewiring an existing switch or receptacle, you should be able to see where the old cables go through the subfloor and into the wall cavity. Be careful not to hit your head on the nails, and wear safety glasses to protect your eyes from falling debris while you drill.

Wiring Through a Finished Wall

Whenever a switch or receptacle is being relocated or added, you may have to route the new wiring through a finished wall. The first step is to deenergize all circuits within the wall. Bring in additional light because deenergizing the circuit may also remove power to all lights in the room you're working in. Then locate the new box, trace around it on the wall and cut out the hole for it (boxes are covered in detail in Chapter 9). A cut-in box is the best choice for remodeling because it can be secured to the drywall easily. It has flaps that attach to the drywall. (If you choose another type of box, it may have to be nailed or screwed to the stud, which means removing a lot of the drywall.) The box should be located about 1 in. from the stud toward the center of the wall cavity. Do not cut the hole for the box so close to the center of the wall cavity that you won't be able to reach the adjacent stud with your drill (see the drawing on p. 108).

If you're running cable horizontally through several studs to a new box, use a utility knife or a keyhole saw to open a 2-in. channel through the wallboard from box to box. Drill through the studs and pull the cable through to the new box.

If you're relocating or adding a box in the wall cavity adjacent to an old box, simply remove the old box from the wall and drill at an angle through the stud to the adjoining wall cavity. Make sure the hole is above or below the location of the old box: If the old box will be staying, it won't fit back into the hole because the new wiring will be in the way.

When a new receptacle is to be installed in a room, and there is one in an adjoining room sharing the same wall cavity, power the new receptacle through the existing one. Turn off power to all circuits in the wall you will be working on. Mark the location for the new box and open up the wall. Be aware that there are wires inside the wall that you don't want to hurt. Do not place the boxes back to back because there probably won't be enough width to the wall. Instead, move the proposed box slightly to one side of the existing one (see the drawing on p. 109).

Wiring Through a Finished Wall

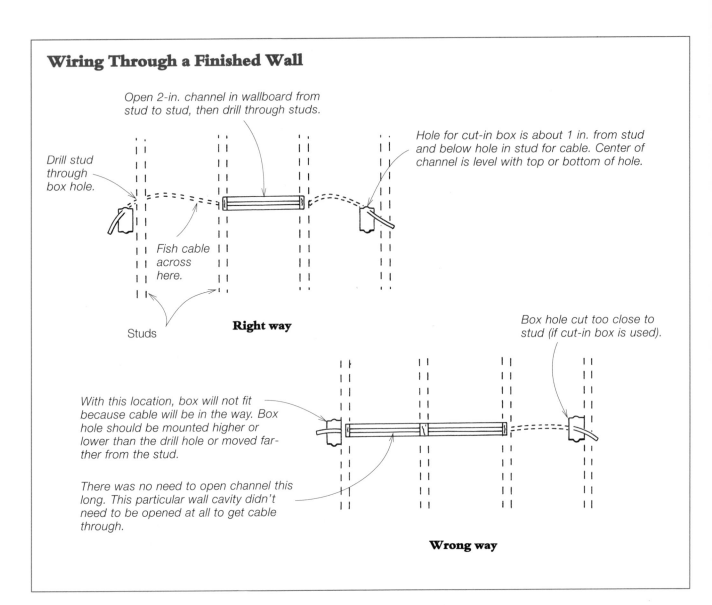

Open 2-in. channel in wallboard from stud to stud, then drill through studs.

Drill stud through box hole.

Hole for cut-in box is about 1 in. from stud and below hole in stud for cable. Center of channel is level with top or bottom of hole.

Fish cable across here.

Studs

Right way

Box hole cut too close to stud (if cut-in box is used).

With this location, box will not fit because cable will be in the way. Box hole should be mounted higher or lower than the drill hole or moved farther from the stud.

There was no need to open channel this long. This particular wall cavity didn't need to be opened at all to get cable through.

Wrong way

Remove the existing receptacle from its box and find which side of the box is nailed to the wall stud. You will have 16 in. center to center from this stud to the next to get the new box in. If the old box is plastic, open up the cable slot with a screwdriver so that the new cable will go through smoothly. If it is metal, loosen the cable clamps and open up one of the knockouts or cable openings with a screwdriver. If you remove a knockout, install a connector to protect the new wire from chafing on the metal. Push a new cable into the old box, giving yourself 6 in. of cable out the box front. Cut the cable to its

approximate length and feed it through the hole for the new box, leaving 6 in. Then install the new box.

Behind Baseboard Molding

You can route cable through the wall behind existing baseboard molding. But only in houses that have molding that can be replaced easily if it gets damaged. Do not try this method on old baseboard that is irreplaceable; it may split or otherwise be damaged as you remove it. This method is most commonly used in situations where the original electrician had left out a feed wire to a switch, or

Wiring Boxes in the Same Wall Cavity

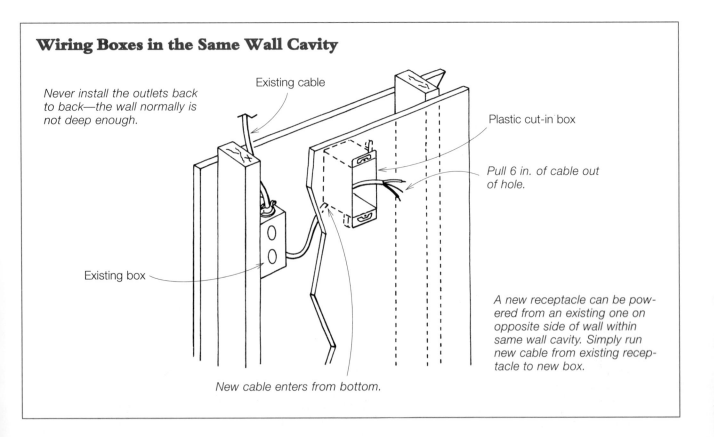

Never install the outlets back to back—the wall normally is not deep enough.

Existing cable

Plastic cut-in box

Pull 6 in. of cable out of hole.

Existing box

A new receptacle can be powered from an existing one on opposite side of wall within same wall cavity. Simply run new cable from existing receptacle to new box.

New cable enters from bottom.

where the owner wants to add a receptacle or fixture after the finished walls are up.

The first step is to deenergize any circuits within the wall. Make a pencil mark along the top of the wall right above the trim. Pop the molding off as carefully as possible. Use a utility knife to cut any paint holding the molding to the wall and use a nail set to drive the trim nails through the wood, which will free the trim board to be removed. If you prefer, you can use a wide chisel or prybar to pull the trim off.

With the trim removed, open up a channel in the wallboard using a keyhole saw or a utility knife. Do not remove any finished wall above the mark. Using a long bit, drill ½-in. or ⅞-in. holes in the studs for the wiring and pull the horizontal section of cable first. Wear safety glasses and be careful not to hit any nails. Run a fish wire down from the proposed box hole, tape the new cable to it and pull both into and out of the hole. Then remove the fish wire and run

the new cable into the box and insert the box into the wall. If you're taking cable into an existing receptacle box, run a fish wire from the bottom of the box to the opened area behind the trim. Tape the new cable to the fish wire and pull both up into the old box (see the top drawing on p. 110).

Around Doors

Routing wiring around a door is normally done when the wall switch has been installed on the wrong side of the door frame, or when a new switch is being added, and power is coming from the receptacle on the opposite side of the door. The procedure also works for windows.

As always, first deenergize the circuit you're working on and any that might be within the wall you're working in. Remove the trim. Then, using a sharp chisel, notch the shims and anything that will prevent the cable from going around the door. Route the cable from the receptacle behind the baseboard,

Routing Cable Behind Baseboard Trim

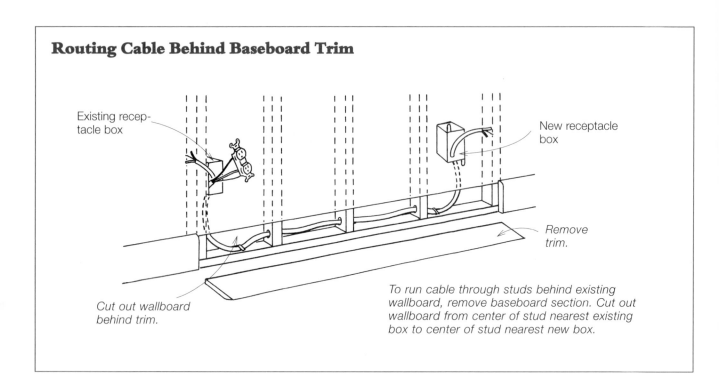

Existing recep-
tacle box

New receptacle
box

Remove
trim.

Cut out wallboard
behind trim.

*To run cable through studs behind existing
wallboard, remove baseboard section. Cut out
wallboard from center of stud nearest existing
box to center of stud nearest new box.*

Routing Cable Around a Door

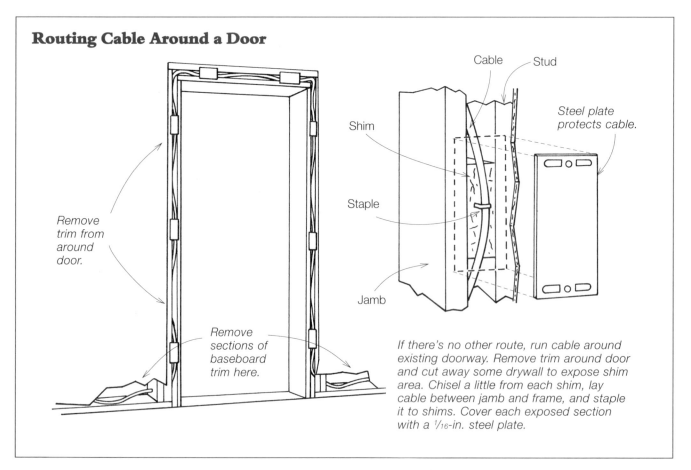

Cable — Stud

Steel plate
protects cable.

Shim

Staple

*Remove
trim from
around
door.*

*Remove
sections of
baseboard
trim here.*

Jamb

*If there's no other route, run cable around
existing doorway. Remove trim around door
and cut away some drywall to expose shim
area. Chisel a little from each shim, lay
cable between jamb and frame, and staple
it to shims. Cover each exposed section
with a 1/16-in. steel plate.*

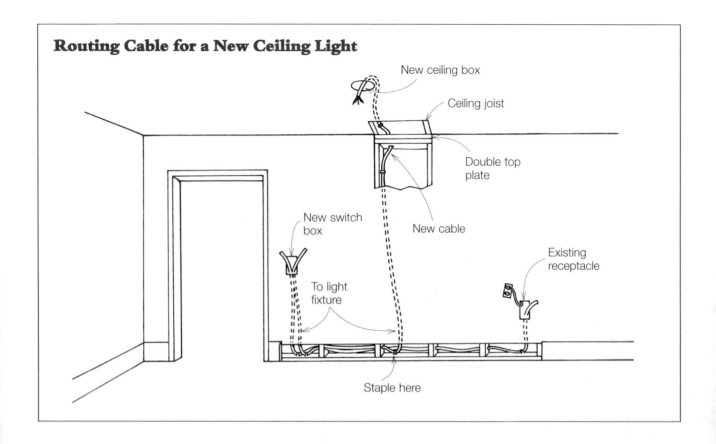

Routing Cable for a New Ceiling Light

New ceiling box

Ceiling joist

Double top plate

New switch box

New cable

To light fixture

Existing receptacle

Staple here

if possible, then up and around the door, squeezing it in around the notched shims. Staple the cable to the shims and protect it where necessary with steel plates (see the bottom drawing on the facing page). If large sections of the cable are exposed, run it through steel conduit at least 1/16 in. thick.

If a switch is to be mounted adjacent to the door, open a hole about 6 in. to 8 in. from the jamb for the switch box. To get the cable from the shim area to the switch box, drill through the door framing from the switch-box side. Insert the cable through the frame and into the box, then install the box.

Through the Ceiling or Attic

Wiring through the ceiling or attic is normally done when the wiring to a ceiling light is being replaced, or when a new ceiling light is being installed. As always, deenergize the circuit to the light and any other circuit in the wall being worked on. If the old light is present, remove it and the old box. If the

light is a new installation, cut out the hole for the fixture box. Be sure to have the box in hand so that you know what size hole to cut (I'll talk more about installing a ceiling light in Chapter 11). It's also important to run the planned wiring parallel to the ceiling joists in this situation so that you don't have to open up the entire ceiling to run the cable through the joists.

Next, remove the baseboard trim, cut a channel in the wallboard and drill through the studs. Next, cut the holes in the wall for the switch box and then cut two slots through the wallboard big enough to get a drill into: one at the top of the wall, and the other in the ceiling. Drill a 1-in. hole through the top plate with the right-angle drill (remember to wear safety glasses). Run the cable from the receptacle to the new switch box to the ceiling fixture using a fish tape or other method, and install the new boxes (see the drawing above).

Chapter 7

FUSES AND CIRCUIT BREAKERS

FUSES

CIRCUIT BREAKERS

Officially, fuses and circuit breakers are called overcurrent protection. Located in fuse boxes and circuit-breaker panels (mains or subpanels), their most common purpose is to protect the wiring from overloads, short circuits and ground faults by interrupting the current flow in the circuit. An overload occurs when a load is placed on a circuit that is more than the rating of the wiring and the overcurrent device. For instance, in a circuit with 20-amp overcurrent protection—through a fuse or circuit breaker—an overload would occur if several appliances were operating at the same time, creating a total current draw exceeding 20 amps. A direct short means that the hot and neutral wires are either directly touching or indirectly touching, and can be created by a nail driven into the wires. A ground fault is a short in which the hot wire touches a ground. For example, a ground fault will occur if the hot wire of a drill touches the drill's grounded metal case.

If the overcurrent protection, whether it be a fuse or circuit breaker, is not in place during an overload, short or ground fault, excessive current flowing through the wire would produce enough heat to damage a wire's jacket and eventually start a fire. There are many different types and classifications of fuses and circuit breakers on today's market. Entire books have been written on this subject alone. But here I will only discuss the common fuses and breakers used in the home.

FUSES

Though today's fuses are an excellent method of overcurrent protection, it wasn't always that way. They, like circuit breakers, had to go through years of product development. Early fuses were simply pieces of small copper wire connected in series in a circuit. Because these wires were small, they had a higher resistance, causing them to overheat and then eventually melt or explode when placed under heavy load. Though such a system worked, it was both a fire hazard and a hazard to workers. Eventually a lead alloy wire was used with contacts on both ends, but under short-circuit conditions, the lead wire exploded just as the copper did. The next innovation was to use zinc as the element and insert the wire in a fiber tube filled with a fire-resistive powder.

Today's fuses are much more sophisticated than those from the early 20th century. Modern fuses blow quickly, which cuts off the current flow in an instant. The problem with old fuses—those from before around 1940—is that it took a long time for the fuse to blow. This delay resulted in massive damage to the circuit as a result of the huge current buildup during a fault. The two most common fuses used today are cartridge and plug fuses. They work basically the same way. An overcurrent will cause the metal element inside a fuse to heat up and eventually melt, interrupting the circuit.

Cartridge Fuses

Cartridge fuses are the oldest type of fuses still used today. These tubular fuses were extremely popular in the early part of this century. If you've got an old house, you'll find cartridge fuses in the fuse box. Made from fiber and designated by UL as Class-H fuses, cartridge fuses are one-time general-purpose fuses, even though reusable cartridges with replaceable links have been developed. One-time means that the entire fuse will need to be replaced if it ever opens the circuit.

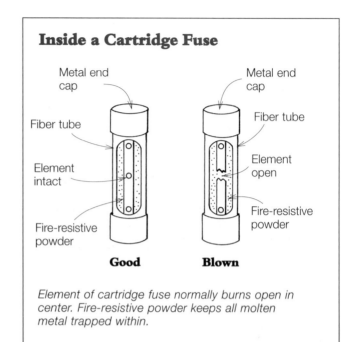

Inside a Cartridge Fuse

Element of cartridge fuse normally burns open in center. Fire-resistive powder keeps all molten metal trapped within.

The two most common types of cartridge fuse are the ferrule and the knife blade. The ferrule type of cartridge fuse protects up to 60 amps. It consists of a fiber tube with metal caps on the both ends that make contact with the circuit (see the drawing above). Inside, across the center of the tube, is a strip of metal that will vaporize when its specific current limits are exceeded. To prevent the arc within the fuse from vaporizing the tube, the strip of metal is surrounded by a fire-resistive powder. A knife-blade cartridge fuse protects above 60 amps. It is basically a ferrule-type of fuse with metal blades on each end that make contact with the circuit. The typical high-end limit is 600 amps, but you can get fuses that are rated up to several thousand amps.

The common problems encountered with cartridge fuses are not with the fuse itself, but with the fuse holder or fuse box. The clips that hold the cartridge fuses sometimes get pulled apart over years of fuse-changing, and the rivets that hold the fuse clips onto the fuse box loosen over time. Both of these conditions result in arcing, burning and overheating of the fuse and sometimes the entire fuse box. Eventually the heat destroys the fuse by

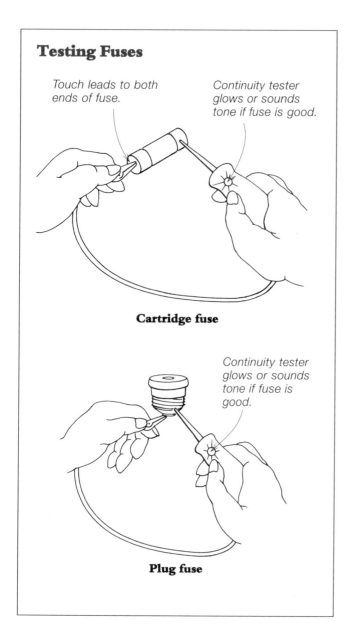

Testing Fuses

Touch leads to both ends of fuse.

Continuity tester glows or sounds tone if fuse is good.

Cartridge fuse

Continuity tester glows or sounds tone if fuse is good.

Plug fuse

it out of the circuit and check its continuity with a continuity tester (also called a fuse tester and circuit tester). A continuity tester verifies that the element of the fuse is intact (meaning there's continuity). Just put a lead on each end of the fuse. The continuity tester will either sound a tone or blink a light if there's continuity, which means that the element is still good (see the drawing at left). If there's no continuity, the element is bad, and the fuse needs to be replaced. You can also test the resistance of the fuse to see if it is good or bad. If the reading is close to 0 ohms, the fuse is good. If the fuse is not good, the meter will read infinity (∞).

Cartridge fuses are no longer common around residences, unless they are used as a fused cutoff switch for an appliance or within the appliance itself. Although they are no longer used a lot in houses, they are still extensively used in commercial, industrial and marine applications.

Plug Fuses

The standard plug fuse (sometimes called a glass fuse) has what is called an Edison, or standard, screw base. They are available in sizes of ½, 1, 2, 3, 5, 6, 8, 10, 15, 20, 25 and 30 amps. These fuses are rated at 125 volts, AC only, and code only permits Edison-based fuses to be used for replacements in existing installations. And even replacements are only allowed if there is no evidence of tampering or overfusing. Fuses of 15 amps and lower have hex-shaped windows, while 20, 25 and 30-amp fuses have round windows (see the photo on the facing page).

The ordinary plug fuse has a mica cover, metal cap, porcelain-type base, fuse wire, or element, center contact and a screw contact (see the top drawing on the facing page). The current goes into the center contact, through the fuse wire, into the screw contact and then to the load. This type of fuse blows very fast and is normally limited to incandescent lighting and other loads that do not have surge currents or temporary overloads. For these types of circuits, time-delay fuses are needed (see the bottom drawing on the facing page). A time-delay fuse looks exactly like an ordinary plug fuse, but the internal

disintegrating the fiber-type shell: I've seen them fall apart in the box and disintegrate as they are pulled out of the clip. Sometimes the overheating is so bad that the fuse box itself starts to melt. No fault can be placed with the fuse or the box; they are simply overworked and old.

It is not always possible simply to look at a cartridge fuse to see if the inside element is intact. To determine whether the cartridge is good or bad, pull

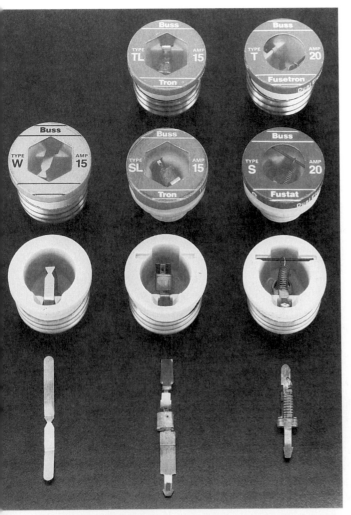

Fuses of 15 amps and lower have hex-shaped windows, while fuses of 20 amps and higher have round windows. The fuse on the left is a general-purpose fuse. The others are time-delay fuses. (Photo courtesy of Bussman, Cooper Industries, Inc.)

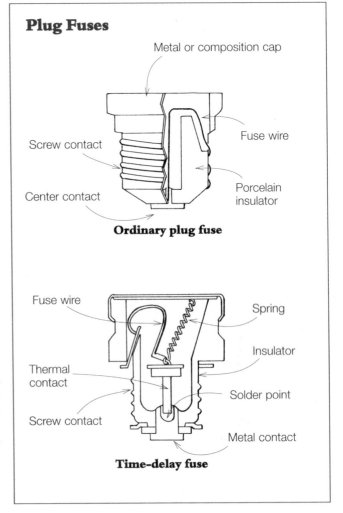

Plug Fuses

Ordinary plug fuse

Metal or composition cap

Screw contact

Center contact

Fuse wire

Porcelain insulator

Time-delay fuse

Fuse wire

Thermal contact

Screw contact

Spring

Insulator

Solder point

Metal contact

construction is totally different. A time-delay fuse has a fuse wire, a spring, a thermal contact and a solder point. Excessive current will cause the thermal contact to heat up. If the excessive current continues long enough, the heat buildup in the thermal contact will melt the solder holding the thermal contact to the metal contact at the bottom of the fuse. At that point the spring will pull the thermal contact out of the solder, breaking the circuit. Time-delay fuses allow five times their rated current for up to six seconds. Fuses that are not time-delayed will normally take 500% of their rated current for approximately a quarter of a second.

Plug fuses can provide protection for appliances around the house at their receptacles. The smallest branch-circuit overcurrent protection for receptacles is normally 15 amps—made to protect the smallest branch-circuit wiring you can use in your house, which is 14-gauge wire. Because plug fuses are made

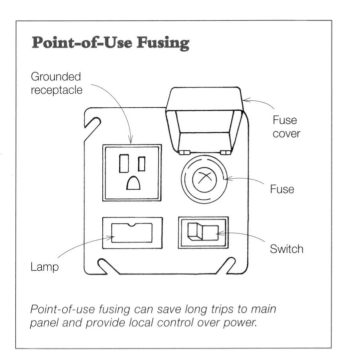

Point-of-Use Fusing

Grounded
receptacle

Fuse
cover

Fuse

Switch

Lamp

Point-of-use fusing can save long trips to main panel and provide local control over power.

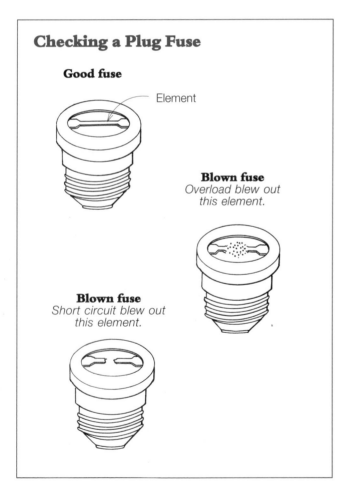

Checking a Plug Fuse

Good fuse

Element

Blown fuse
Overload blew out this element.

Blown fuse
Short circuit blew out this element.

in a wide variety of sizes, all the way down to ½ amp, they are ideal for motor protection. If you have a plug-in appliance or motor, it can be fused at the receptacle at considerably less or even equal to the branch circuit breaker (see the top drawing at left). An example of its use would be a table saw that, when stalling out, always trips the breaker. The circuit breaker may be far from the table saw, perhaps even on a different floor or in a different building. Fusing at the receptacle saves long trips to the fuse panel or circuit breaker and provides local control over the power. Another purpose for point-of-use protection is to fuse an appliance at considerably less than the 15- or 20-amp protection at the main panel. Let's say you have an appliance that pulls only 1 amp. Without a point-of-use fuse, the breaker at the main panel will open the circuit during an appliance malfunction only when it exceeds 15 or 20 amps; but by that time, the motor or appliance would be ruined. In this situation, the breaker in the main panel protects the wiring, but the fuse protects the appliance.

The problem with Edison-based fuses is that they are all the same physical size and have the same thread design, so they are easily confused—a 15-amp fuse can be unscrewed and replaced with a 30-amp. To counter this, S-type Edison-based adapters were developed, which, with matching S-type fuses, screw into the base threads of a standard household fuse box. These adapters—now required by code—prevent a person from inserting the wrong-size fuse into the base by varying the base threads. There are three classifications for the screw-base adapters: 0 to 15 amp, 16 to 20 amp, and 21 to 30 amp, all of which match associated adapters on the fuses to prevent any mixups. S-type fuses are distinguished by their nonmetallic threads.

To see if a plug fuse is good or bad, just look through the window (see the bottom drawing at left). If the element is intact, the fuse is good. If the element is not intact, you can find out what caused it to blow simply by looking at it. If the element is burned in the middle, with its ends intact, the fuse was blown from an overload. If the element appears exploded in the middle, and the window is darkened as well, a short occurred on the circuit. If in doubt about

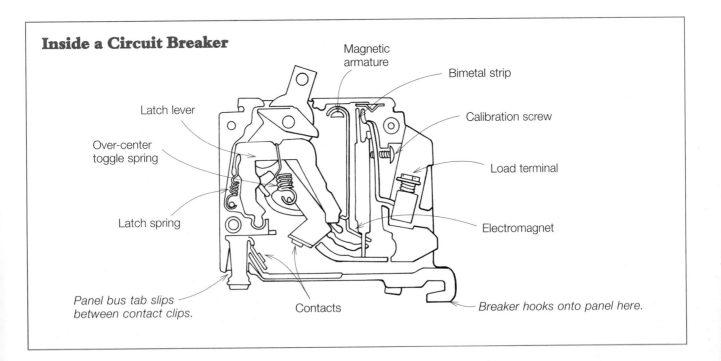

Inside a Circuit Breaker

Magnetic armature

Bimetal strip

Calibration screw

Load terminal

Electromagnet

Latch lever

Over-center toggle spring

Latch spring

Panel bus tab slips between contact clips.

Contacts

Breaker hooks onto panel here.

whether a fuse is good or bad, it can be checked either by testing its continuity or resistance (see the drawing on p. 114). Plug fuses are covered in Section 240E of the NEC.

Old–Style Main Fuse Boxes

Old 60- and 100-amp fuse boxes are still providing overcurrent protection in many aged homes throughout the country. These elderly fuse boxes are easily recognized. Most have four plug fuses across the bottom for four general-purpose receptacle circuits. At the top left of the box is usually the main circuit, and to the top right, the stove circuit—both protected with cartridge fuses (many were wired backward, with the main on the right and the stove on the left). The water heater was wired via an internal tap off the main, and many times this tap gave no overcurrent protection for the wire going to the water heater.

These boxes were, for the most part, satisfactory for the loads that were placed on them when they were installed. But as the years went by, circuits and appliances were added—overloading the circuits—and eventually fuses started to blow. Instead of

upgrading the service, many people simply installed larger-rated fuses, not realizing the disaster that could result. Wiring that needed 15- and 20-amp protection was fused at 30, allowing excessive current to heat up the wiring and kill the fuse box or start a fire. If you live in an old house, or are working on one, check to see if there's an old 60-amp fuse box providing overcurrent protection for the house. If you see 30-amp plug fuses in the box, with 12- or 14-gauge cables running to their respective branch-circuit screws, the service should be upgraded immediately. **Warning!** Never replace the original fuses with higher-rated fuses to prevent the fuses from blowing. Likewise, never modify a fuse in any way, such as inserting pennies or wrapping aluminum foil around cartridge fuses. Doing so will always result in disaster.

CIRCUIT BREAKERS

Circuit breakers—officially called molded-case circuit breakers, or MCCBs—are the most common type of overcurrent protection used in homes (see the drawing above). Simply put, a circuit breaker will sense and measure any amount of overcurrent in a circuit. If the measured overcurrent exceeds a predetermined amount, the circuit breaker will trip off.

Most circuit breakers use a two-part system to protect the circuit: one for mild overloads, and the other for severe short circuits. In a mild overload, a bimetal strip, called the thermal trip, heats up and bends backward, eventually tripping the breaker. For severe short circuits, an electromagnet gives an assist to the bimetal strip, pulling it back even faster for an almost instantaneous response to open the circuit. In general, the higher the current, the less time it takes for the breaker to trip.

When the breaker does trip, it will open to a position almost halfway between on and off. After you have identified the problem on the circuit and remedied it, you can turn the breaker back on. (You cannot hold a breaker on against an excessive load.) To turn the breaker on again, you must turn the breaker to the full off position and then back to on.

Wiring a Circuit Breaker

Hooking up a circuit branch to a breaker is fairly simple, whether copper or aluminum wire is used. First, strip off just enough insulation to slip the wire into the breaker. Don't strip off too much insulation, because if over ¼ in. of bare wire is showing outside the circuit breaker, the wire could short out on another wire. Be sure to get the screw tight and have only one wire per breaker.

When aluminum wire is to be connected to a breaker, follow these instructions: Strip the insulation completely from the end of the wire, without nicking or cutting the wire. Wire-brush the exposed conductor strands to remove any oxidation from the wire (the wire should be shiny). Apply a coat of an approved antioxidant, and insert the conductor into the breaker, making sure that all strands are contained. Then tighten the screw to the specified torque indicated on the breaker (normally around 20 in. lb.).

Although you may be tempted to do so, never trim a heavy-gauge wire to reduce its size to fit into a breaker. Instead, splice the wire to a smaller gauge wire that will fit into the breaker and under its screws.

Different manufacturers attach their breakers to the panel in different ways, so it's important to follow the directions from the manufacturer. The most common method of installation is to first hook the wiring side of the breaker under a panel lip, or notch. Then push the breaker down onto the bus tab—do not touch the hot bus! You should feel a solid stopping point as the breaker seats.

1. Make sure breaker is off.

2. Swing breaker into panel notch.

3. Press breaker into tab and fully down into panel.

Do not touch hot bus!

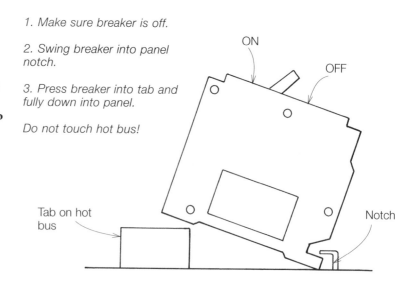

Inserting a typical circuit breaker

A breaker is also designed to be ambient-compensated. As the air around the breaker heats up, the breaker senses this heat and derates itself accordingly, meaning that a 15-amp breaker could lower its rating to 14 amps and lower if enough heat is generated in the panel.

Most breakers for the home are designed for wire insulation maximum temperatures of 60°C and 75°C and for copper and aluminum conductors. The temperature rating and CU/AL should be printed on the breaker (see the sidebar on the facing page to find out how to wire a breaker).

In addition to temperature, every circuit breaker has an ampere rating, a voltage rating and a short-circuit interrupting rating. The ampere rating is printed on the breaker's handle and indicates the maximum current the breaker can carry without tripping. Standard residential breakers are rated for both voltages available in the panel—120/240 volts. A good general rule to remember is that the breaker's maximum voltage rating must equal the maximum voltage that will be applied to it. The short-circuit interrupting rating is based on the maximum amount of fault current available from the transformer outside the residence. Simply put, if the transformer can generate 10,000 amps of current, the breakers in a residence should be rated to at least 10,000 amps. Residential breakers have ratings from 10,000 amps to 65,000 amps. However, for a single-family dwelling, the available fault current is rarely above 10,000 amps.

Manufacturers design circuit breakers specifically for their own panels, and you are required to follow the instructions given by the manufacturer. But many breaker designs are universal and can be installed on panels made by another manufacturer. You are not allowed to insert one manufacturer's breaker into another manufacturer's panel unless the panel's manufacturer approves that breaker and says so on the panel. If you try to use a breaker made by company A on a panel made by company B, without approval from company B written on the panel, an electrical inspector has the right to reject the job because the manufacturer's instructions were not followed. Not only that, but the warranty on the panel will also be void.

Single-Pole Breakers

Single-pole breakers, rated at 120 or 120/240 volts AC, control current on loads that use only one leg (120 volts) of the available 240 volts in the panel. Standard single-pole breakers are made in 15-, 20-, 25-, 30-, 40- and 50-amp ratings, with 15 and 20 being the most common in a residence. Single-pole breakers are available in three varieties (see the drawings on p. 120): full and dual (1 in. wide), with some manufacturers making half-size (½ in. wide). Two half-size breakers grouped together create a dual breaker, which allows you to get two branch circuits on one slot in a panel. But the panel must be designed to accommodate the dual or half-size breakers—some only allow full-size breakers, others, a fixed number of dual and half-size breakers, and a few will allow the panel to be totally filled with the dual or half-size breakers.

I prefer to design the system so that neither dual nor half-size breakers are needed: I prefer to use only full-size breakers with panels designed only for full-size breakers. But when I cannot get around it, I'll choose dual breakers that are enclosed in one full-size case, which makes handling and installing them easier— the half-size breakers are about as thin as pancakes. Doubling up breakers can create a few drawbacks: First, the wiring inside the panel winds up looking like a bird's nest with no room to work, but more important, it's easy to violate the code by doubling up breakers.

The violation occurs when dual breakers are installed where they are not meant to be. I said before that only panels designed for dual breakers are allowed to use them. The buses on such panels have a few tabs

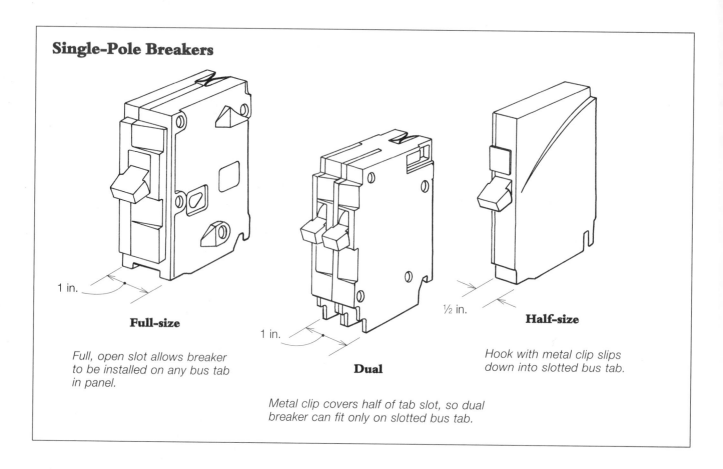

Single-Pole Breakers

Full-size

1 in.

Full, open slot allows breaker to be installed on any bus tab in panel.

Dual

1 in.

Metal clip covers half of tab slot, so dual breaker can fit only on slotted bus tab.

Half-size

½ in.

Hook with metal clip slips down into slotted bus tab.

at the bottom of the panel that can accommodate dual breakers. Such a tab has a slit in it. A special clip on the dual breaker fits into this slit, which prevents it from fitting onto a standard bus tab. However, if you know how, the clip can be removed, allowing the breaker to fit onto a standard bus tab, which is a code violation. Each panel is allowed only a certain number of breakers within it. By removing the clip and inserting dual breakers where they are not supposed to be, you can literally double the amount of breakers in the panel. Bill Goode, a past electrical inspector for Roanoke County, VA, told me of an inspection he did, where clips from these half-size breakers were littered all over the floor in front of the panel. Looking into the panel, Bill could see an overabundance of dual breakers. Needless to say, the inspection failed.

Here is a typical situation that leads to a violation. After determining that you need a 200-amp panel, you go shopping for one. You find one for $125 and another for $175. Not paying attention to panel design and quality, you simply go for the lowest price, thinking you saved $50—after all, you say, they're both 200-amp panels. You install the panel and start the wiring. Before long, the panel is totally full of breakers, and you have more circuits to install. So, what to do? You can pull out all the wiring, remove the panel and go buy the $175 model. (Now you know it cost more because it could hold a much larger complement of breakers. I've seen new houses go up with boxes that held only 24 circuits, which I think is nothing for a new house.) However, removing the panel takes a lot of work, and you will have all that extra expense of buying another panel. So you check to see if anybody is looking. If not, you

Double-Pole Breaker

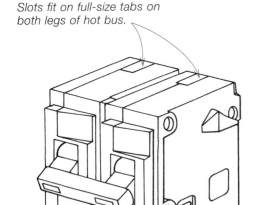

Slots fit on full-size tabs on both legs of hot bus.

2 in.

Typical double-pole breaker fits across both legs of hot bus and is used for circuits that power 240-volt and 120/240-volt appliances. It consists of two single-pole breakers sharing common trip.

Quad Breaker

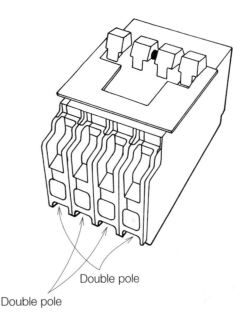

Double pole

Double pole

Quad breaker can obtain two double-pole circuits in same physical space as one standard double-pole breaker. It can power two 240-volt appliances off two slots in panel. A quad can be dangerous because it is possible to shut off only one side of breaker and leave one side powered.

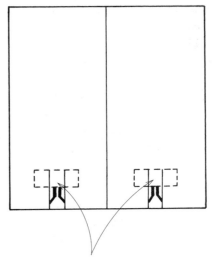

Internal metal clips allow quad to be attached only to tabs with slots.

insert dual breakers—where they're not supposed to be—so that you can get two breakers in the same space as one full-size breaker. Naughty, naughty. Please remember, not only is this a code violation, but it also voids the warranty of the panel.

Double-Pole Breakers

Double-pole breakers, rated at 120/240 volts, control the current on loads that use both legs of the available 240 volts at the panel. They are available from 15, 20 and 25 amps, all the way up to 200 amps. The 15- and 20-amp breakers are normally used for baseboard heaters and pumps; 30 amps are

Limitations of Fuses and Circuit Breakers

It's important to note that just because a wiring system has overcurrent protection—whether it be from fuses or circuit breakers—does not mean that the system is completely safe to people and appliances. Overcurrent protection is supposed to protect wiring and appliances from shorts and surges, but there are instances in which fuses and breakers will provide no protection at all. In general, fuses and breakers will not protect you against shocks, overheating of appliances, low-level faults, surges and fires within an appliance.

Electric shock
Most overcurrent devices will not protect you from a minor shock or even being electrocuted. Overcurrent devices protect circuit wiring from overheating due to excessive current; they have nothing to do with protecting people from shocks.

Overheating
If current flow through a device is impeded by a large amount of resistance, such as a loose connection, significant heat can be created. Overheating can also result from excessive current flowing through a device. Such conditions can completely burn or melt the wire insulation at the point of connection without tripping the overcurrent device.

I once did a service call where the receptacle was not just overheating: It was also glowing cherry red. Turns out there was a loose connection behind the receptacle. The same situation could have been created by a loose splice or a corroded terminal. On another service call, I caused a small explosion when I let air into an arcing, corroded splice box.

Low-level faults
A low-level fault occurs when a hot wire makes a connection to the neutral or ground, creating a low current flow that is not high

enough to trip a breaker or blow a fuse. All it would take to create a low-level fault is to somehow have the current limited to an amount below the breaker trip amount.

A few things can happen if a low-level fault occurs. A person could be electrocuted from the neutral or ground without a load being connected, or a fire could start. Theoretically, this could happen from both line and extension cords that crack and become frayed, especially those that are overhanging baseboard heaters. They arc and burn or they create a semiconductor bridge, allowing current to flow to a noncurrent–carrying conductor. One time I was working on a baseboard heater in a local church. Due to an internal fault in the heater, when I checked the current on both legs of the 240-volt circuit with a clamp-on ammeter, I noticed that one was pulling several amps less than the other. When I placed the ammeter

used for water heaters, dryers and heat pumps; 40- to 50-amp breakers are for stoves; and breakers larger than 50 amps are used for electric heat. Some of the larger amperage breakers may take up to six panel spaces. Double-pole breakers are normally two single-pole breakers with one handle and a shared internal trip mechanism (see the drawing on p. 121). A different type of double-pole breaker, made by Crouse-Hinds, occupies only 1 in. of panel space. Like other half-size breakers, it will only fit onto a slotted tab.

Just as in single-pole breakers, double-pole breakers come in two-in-one units. If you have only two slots available in the panel, and you need four slots for two, 240-volt appliances, you'll need to install this type of breaker, but only if the panel is designed for slotted tabs (see the drawing on p. 121). This type of setup is called a quad, and I don't like it. One 240-volt circuit is protected by the breaker in the center, and the other circuit is protected by the outer breakers, with each outside breaker protecting half the circuit. The outside breakers are ganged together around the grounding wire, I found the missing current: 6 amps were flowing on the grounding wire. Even though the unit was grounded, anybody could have been electrocuted simply by touching the metal of the baseboard heater. The solution was to replace the heater.

Surges

Surges are high-voltage and current pulses generated from both appliances and lightning. Fuses and circuit breakers cannot protect against these surges because the pulses only last for a few microseconds, and the overcurrent protection simply cannot act quickly enough to stop them.

Surges created by appliances don't pose a significant threat to people, but they can damage electronic equipment. However, surges generated by lightning can cause significant damage to both property and people, and lightning rods won't stop them.

Voltage surges generated by lightning can enter a house in two ways: Through the power lines, and then into the house through the service wires, or through saturated induction. Surges through the power lines can be stopped with surge arresters installed at the main panel (see pp. 76-77). But saturated induction is another story. Lightning produces massive amounts of magnetic waves that simply saturate the house and the surrounding area. Anything that is a conductor will have a pulse induced within it. These pulses are created on both sides of the overcurrent protection and even into the grounding system itself. If you're touching a microwave, and your bare feet are on a concrete floor, you could get a shock. If the grounding system is very good, the shock will feel like a needle jab—the same as a GFCI jolt. But if the grounding system is bad (high resistance), the shock could be fatal.

Fires from within the appliance

If your appliance malfunctions internally and starts to burn, the overcurrent protection probably will not trip, which is why many houses burn down. I've seen a timer in a clothes washer catch fire, and I've seen televisions, boom-boxes and a computer go up in smoke. In none of these cases did the breaker trip. Until the insulation melts, and a ground fault or short occurs, there will not be enough current to trip the breaker.

via a long angled bar, making it possible to turn only one half of the double-pole circuit off at a time. This in turn makes it possible to leave the other half on by accident to electrocute someone. This type of breaker can confuse the homeowner as well, and so, in my opinion, it should be banned. Quads can come with different current ratings for each of the attached double-pole breakers; e.g., a 20-amp double-pole breaker can be paired with 30-amp double pole.

Chapter 8

GROUND-FAULT CIRCUIT INTERRUPTERS

As a professional, I've used several tools over the years that malfunctioned and placed me into the electrical path. One morning I was drilling some treated posts outside so that I could attach electrical boxes to them. Dew was on the ground, my shoes were wet, and I was using an old, metal-case drill. About halfway into drilling a post for the first electrical box, the hot wire in the drill shorted over to the metal case and into the handle. Current started flowing into my hand, contracting the muscles and making my grip tighter on the tool I desperately needed to release. The current flow went through my arm, my body and heart, and finally through my legs and into the ground. Luckily, the drill was plugged into an outdoor receptacle that was protected by a ground-fault circuit interrupter (GFCI). So instead of being killed, I felt just a sharp jab, like a pin prick, and then the GFCI opened the circuit, stopping the flow of current. I literally owe my life to the person who developed the GFCI. (Thank you, whoever you are.) My story is not unique: Thousands of people like myself owe their lives to the use of GFCIs.

GFCIs are inexpensive electrical devices, easy to install and available as either receptacles or circuit breakers. Just as overcurrent devices protect circuits, GFCIs protect lives. The old movie scene of a radio dropping into the tub to electrocute the ill-fated bather could not happen in a bathroom properly wired with GFCI protection. This chapter will illustrate how GFCIs work, how to test them, and more important, where and how to install them properly. (The NEC talks about GFCIs in sections 210-7d, 210-8, 305-6 and 555-3.)

HOW A GFCI WORKS

General-purpose, 120-volt household circuits have current flowing to and from the load on two insulated wires. Power is brought to the load on the black wire, flows through the load and then returns via the white. A GFCI compares the current flowing to the load with the current coming from the load. The current, or amperage, should be equal (see the photo at right). If there's a difference, the electrons must be flowing somewhere other than the load (such as through you to ground), and the GFCI will open the circuit (see the drawing below). This current can be as low as .006 amp and doesn't need a grounding wire to work.

The black and white wires pass through a coil (bottom of photo) in the GFCI that compares the current flow in each. If the currents are not equal, the circuit is broken.

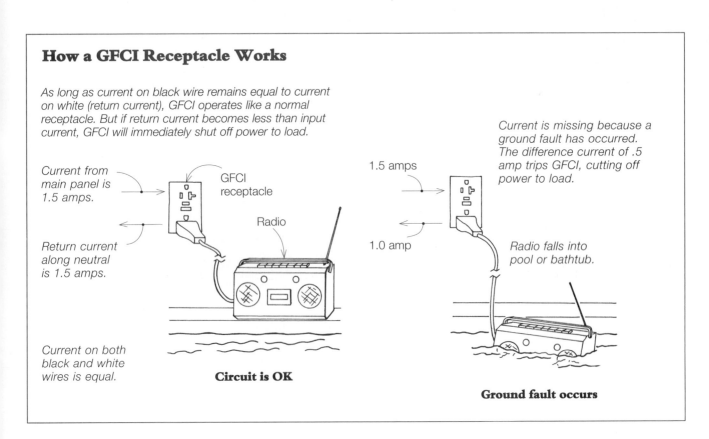

How a GFCI Receptacle Works

As long as current on black wire remains equal to current on white (return current), GFCI operates like a normal receptacle. But if return current becomes less than input current, GFCI will immediately shut off power to load.

Current from main panel is 1.5 amps.

GFCI receptacle

Radio

Return current along neutral is 1.5 amps.

Current on both black and white wires is equal.

Circuit is OK

Current is missing because a ground fault has occurred. The difference current of .5 amp trips GFCI, cutting off power to load.

1.5 amps

1.0 amp

Radio falls into pool or bathtub.

Ground fault occurs

A GFCI Can be Fooled

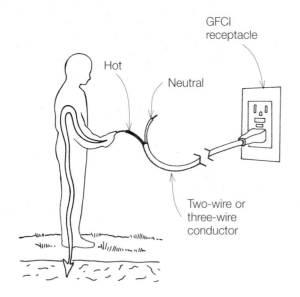

Ground-fault situation

Current flows from hot wire into hand, down through body and into ground. GFCI senses difference current and trips.

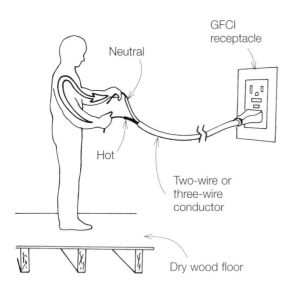

No ground fault

Current flows from black wire through body and out of other hand into neutral wire, with no current flowing into ground. No ground fault occurs, and no current is missing, so GFCI will not trip.

The GFCI opens the circuit in ¹⁄₂₅ to ¹⁄₃₀ of a second, 25 to 30 times faster than a heartbeat. You will still receive a mild shock, which will feel like a pin prick. But don't panic. If you receive a shock while on a ladder, freeze. Jerking your arms or legs wildly could cause you to lose your balance and fall off the ladder, harming you more than the shock itself.

As good as GFCIs are, they can be fooled. As long as the current flowing through the black and white wires is equal, the GFCI operates as if everything is okay. However, if you're standing on an insulator or nonconductive surface, like a dry board, and place your body between the black and white wires, the electricity will flow out the black wire, through your body, and back through the white wire. No current will leak through you to ground, so a ground fault won't exist (see the drawing at left). And a GFCI can't tell the difference between you and an appliance, so your heart will be sent into ventricular fibrillation, a wildly erratic rhythm, and you'll be dead in a few minutes. It would be as if there were no GFCI protection on the circuit at all.

A circuit breaker would not have helped. The reason is because breakers for general-purpose receptacles trip only when the current exceeds 15 or 20 amps—2,500 times more than it takes to produce death. Breakers, or even fuses, are designed to protect the wiring within the household against excessive current. They are not designed for life protection.

TYPES OF GFCIS

The two most common types of GFCIs used around the home are the receptacle and circuit-breaker types (see the photo on the facing page). The circuit-breaker type fits into the main control panel just like a standard breaker and will, if wired correctly, protect all the receptacles on the circuit. Receptacle-type GFCIs fit into a standard receptacle box and will protect any receptacle wired downstream.

The circuit-breaker type of GFCI is more expensive than the receptacle type, and it is less convenient:

The receptacle GFCI can be reset at its point of use while a circuit-breaker type requires you to go to the main panel. I recommend using receptacle GFCIs inside the house and circuit-breaker types for the outside circuits (remember, the NEC requires all outside outlets to be GFCI-protected). The circuit-breaker GFCIs simply last longer than outdoor receptacle GFCIs because they are located inside the residence, not exposed to the weather.

The most common complaint concerning GFCI receptacles is their short life span (they should be tested regularly—see pp. 132-133). Though some do last a long time, many are bad upon installation, and others last only a few years. Residential-grade GFCIs are the cheapest and the most unreliable. Personally, I have had so much trouble with residential-grade GFCIs that I no longer install them. The two I installed on my deck a few years ago no longer have their life-protection system working, and I've had many others fail after installation. Now I install only commercial-grade units, which are of much better quality for just a few dollars more.

GFCI receptacles are available in 15- and 20-amp configurations. GFCI breakers are available from most manufacturers up to 30 amps, and a few manufacturers make breakers up to 60 amps. Breakers rated above 60 amps are hard to find. GFCIs should be sized the same way you would a regular receptacle or breaker. A 20-amp GFCI circuit breaker, for instance, goes with a circuit containing 12-gauge wire. Do not install a 20-amp unit on a branch circuit that uses 14-gauge wire (see the drawing at right).

An assortment of GFCI receptacles. At top right is a 20-amp GFCI-protected circuit breaker.

20-Amp GFCI Receptacle

Horizontal slot off wide prong indicates receptacle is 20 amps. Never use a 20-amp GFCI on circuits with 14-gauge wire. A 20-amp receptacle will allow a 20-amp load to be connected.

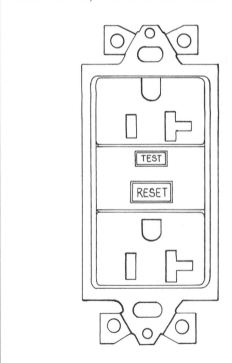

Portable GFCI Protection

All job sites now require GFCI protection, yet in a renovation or addition, the house wiring may not have any GFCI-protected circuitry. To that end, GFCIs are now available in many portable configurations. Although these GFCIs are used often by professionals, homeowners also can take advantage of them.

One of the most handy is the plug-in GFCI adapter, which allows you to provide GFCI protection in any receptacle in the house, and you can pull it out and take it with you when you move (an especially useful item for apartment-dwellers). In-line GFCI adapters are available, which can be attached to extension cords, tools or appliances to provide ground-fault protection, as well as GFCI extension cords. GFCI-protected outlet plugs, which look like typical power strips, provide you with several ground-fault protected outlets and surge protection.

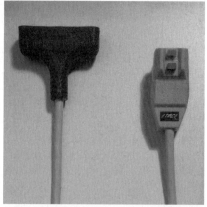

These outlet strips provide both GFCI and surge protection.

GFCI plug-in adapters (single and triple outlets shown) provide protection at the outlet.

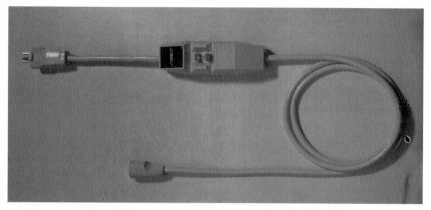

An in-line GFCI cord set can be attached to tools, appliances or extension cords. (Photos this page courtesy of Technology Research Corp.)

If an outdoor cord will be plugged in and left, a box or lid suitable for wet locations while in use is required. These covers can be completely closed while the cord is plugged into the outlet. Both flush-mount and recessed models are available, as well as covers for switches. (Photo courtesy of TayMac Corp.)

WHERE GROUND-FAULT PROTECTION IS REQUIRED

GFCIs are very sensitive—they will trip with a difference current as low as .006 amp—so they should not be installed where life or property would be in jeopardy should power be cut off. For example, freezers, refrigerators, sump pumps and medical equipment should not be on GFCI-protected circuits. In addition, lights—unless you have a good reason—should never be on a GFCI-protected circuit. If the GFCI trips, you literally could be left in the dark, which could be dangerous, especially if you're trying to find your way out of a dark, wet bathroom. Also, GFCIs cannot be used for ranges, ovens, cooking appliances and clothes dryers whose grounded neutrals are connected to the frame of the appliance.

For life protection, the NEC requires GFCIs in a few locations in a house: in bathrooms, kitchens, garages, crawlspaces, unfinished basements and even outdoor locations, such as decks, porches and outbuildings (and don't forget the job site, as mentioned before).

In wet locations, such as kitchen countertops, all receptacles must be GFCI-protected. Most appliance cords are 6 ft. long, so any appliance within 6 ft. of a wet-bar sink can fall into the water. Because bathrooms are almost always wet, all receptacles in the bathroom need GFCI protection, whether located in a light, in a medicine cabinet or in a wall. The overhead bathroom fan or light—or combination fan/light—needs GFCI protection when it's mounted in the bath or shower enclosure. In addition, the unit will need to be listed for wet locations.

All outside receptacles, including those in outbuildings, need GFCI protection. If the GFCI is to be used outdoors, and its load is to be plugged into the receptacle day and night, rain or shine, the outlet must be covered. Typical examples are Christmas-tree lights, fence chargers and low-voltage transformers. The receptacle cover must be suitable for wet locations while the appliance is in use (see the photo above). These covers look like bubbles on the front of receptacles. You cannot use a cover with a metal, spring-latched door. This type of door is

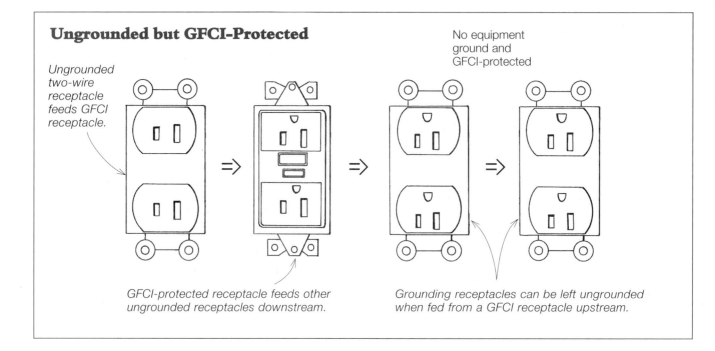

Ungrounded but GFCI-Protected

Ungrounded two-wire receptacle feeds GFCI receptacle.

No equipment ground and GFCI-protected

GFCI-protected receptacle feeds other ungrounded receptacles downstream.

Grounding receptacles can be left ungrounded when fed from a GFCI receptacle upstream.

used simply to cover the outlet when it's not in use. Even though these receptacles are covered, moisture still creeps in and destroys the mechanism that makes the GFCI work. It will still provide power, but the life-protection will be lost. That's why I recommend using circuit-breaker GFCIs for permanent outdoor installations.

GFCI protection is required in any general-purpose receptacle in a garage; in a crawlspace at or below grade level; in an unfinished basement; in a boathouse, dock or seawall; and in areas around swimming pools, spas and hot tubs. Spas, hot tubs and swimming pools get quite complicated when it comes to GFCI protection, because of the obvious danger involved, and should not be handled by an amateur. These areas should be looked at and wired by a licensed master electrician.

Any receptacle, whether two-wire or three-wire, that goes bad and needs to be changed and is located in an area currently requiring GFCI protection, must be replaced with a GFCI receptacle. For example, if you're changing a receptacle in a bathroom or along a kitchen countertop, the replacement must be GFCI

protected. Also, grounded receptacles can be used ungrounded if fed from a GFCI receptacle (see the drawing above).

Exceptions

There are exceptions to the requirements. For example, you may have a corded appliance in a garage or basement that will always be in the same location—such as a freezer or refrigerator—and you don't want to worry about a GFCI tripping often and losing power. In that case simply install a dedicated single receptacle for the appliance (or a duplex outlet if there are two appliances) that's not GFCI-protected, where only that appliance can be plugged into it. Another exception is receptacles that are not readily accessible, such as a garage-door opener. But remember, just because GFCIs are not required, like for the garage-door opener, doesn't mean you can't put one there. You can if you want to. I've had my garage-door opener on a GFCI-protected circuit for many years with no trouble. I installed the GFCI so that it will trip (and it sometimes does) if lightning surges come down the line. Lightning loves to destroy garage-door openers.

WIRING RECEPTACLE GFCIS

It is imperative that receptacle and circuit-breaker GFCIs get wired properly because the function of these units is to save lives. And with so many wires involved, it's easy to connect one wrong. GFCIs have both load and line connections—four screws and wires to figure out, plus the ground wire. Sometimes the load connections are on the right, sometimes on the left, and other times at top or bottom. There is no way to tell any of the connections apart without reading the print on the back of the GFCI. And the print is normally hard to read. Some GFCI receptacles will provide power even if they are wired backward (power from the service panel is wired to the load side instead of to the line side). If the downstream receptacles are wired from the line side of the GFCI, they may not work correctly: They may not be able to sense a current imbalance. Worse, if the GFCI is wired wrong, it may still test normally (see the drawing at right). To wire a GFCI receptacle, follow these steps:

1. Locate the terminals LINE HOT and LINE NEUTRAL. Connect the feeder wires from the main panel (black to hot, white to neutral) here to bring power into the GFCI. Once connected, protected power will be available at the GFCI itself and its load terminals. Unprotected power is taken off the two feeder wires before they attach to the receptacle line terminals. Never put two wires under any screw; instead, use wire nuts and pigtail off the splice to the GFCI (for more on splicing, see pp. 152-154).

2. Identify the terminals for LOAD HOT and LOAD NEUTRAL. This is outgoing protected power, if you want it. If you have no need for any more protected receptacles in the circuit, nothing will connect to these two terminals. However, any receptacle that uses these two terminals for its feeder will have ground-fault protection provided by the GFCI receptacle. If used, all protected receptacles should be marked on the receptacle cover as "GFCI protected." Little stickers are normally provided with the GFCI receptacle just for this purpose.

Wiring a GFCI Receptacle

GFCI receptacle can provide ground-fault protection to regular receptacles downstream, as long as the wiring isn't reversed. Wires from main panel should be attached to line side; wires feeding receptacles downstream should be attached to load side.

3. Connect the grounding wire. Though the grounding wire should always be connected if a grounding circuit exists, a GFCI doesn't need a ground wire to work. All it needs to cut off the circuit is a current imbalance from hot to neutral. In fact, GFCIs can replace old two-wire receptacles (see the drawing on the facing page).

WIRING CIRCUIT-BREAKER GFCIS

A GFCI circuit breaker protects wiring just as a standard breaker does—tripping when the current exceeds the breaker's rating. But it also provides GFCI protection, tripping when there's current missing in the circuit. Before installing the breaker, make sure it matches the gauge of the branch's wiring: a 15-amp unit with 14-gauge wire, a 20-amp unit with 12-gauge wire.

Wiring a GFCI Circuit Breaker

To load hot, normally black or red

GFCI breaker

To load neutral, normally white

White with black stripe

Neutral bus

GFCI circuit breaker attaches to both hot and neutral wires from branch circuit. If breaker has just two terminals (hot and neutral), be sure you don't reverse wiring. Black-and-white wire connects to neutral bus and is breaker's 0-volt reference.

Wiring a GFCI breaker is fairly simple, but it still requires care because the wiring can easily be reversed. A standard GFCI breaker has a white wire with black stripes attached to it. This wire, normally coiled like a pig's tail, connects to the neutral bus and is the breaker's 0-volt reference. The breaker will also have another solid white wire attached to it, which connects to the neutral from the load. The black (hot) to the load connects to the hot terminal on the breaker (see the drawing below left). However, some breakers simply have screw terminals for the neutral and hot wires. If so, take a close look at the two screw terminals—the writing will be hard to read. One will read neutral and will receive the circuit's white wire; the other will read hot and will receive the circuit's hot wire. Do not reverse these two wires! If reversed, the circuit may test normally using the GFCI's test button. It will even trip using GFCI push-button testers on the downstream receptacles but may not provide ground-fault protection. The only way you will know there's a problem is by using a plug-in tester on one of the downstream receptacles.

TEST GFCIS FREQUENTLY

All GFCIs should be tested at least once a month with their own test button, and the receptacles they protect should be tested with a plug-in GFCI tester. The test button places a current imbalance on the circuit to see if the GFCI will open the circuit.

When the test button opens the circuit, a receptacle GFCI will open both the neutral and line terminals. So the GFCI will work, even if the hot and neutral wires are reversed. However, the GFCI breaker only opens the hot line—the neutral stays intact. If the breaker has been wired backward, the intact line is now the hot line, and current can still go to the load. After the test, press the reset button on the receptacle to ready the GFCI. To reset a GFCI breaker, turn the breaker to full off and then on again.

GFCIs are not toys. I mention this because one of my customers had a child who loved to press the test and reset buttons on a GFCI receptacle I had just installed. She liked to hear the click as the GFCI

This simple circuit tester plugs into a receptacle and creates a current imbalance. If the GFCI is working, the circuit should trip.

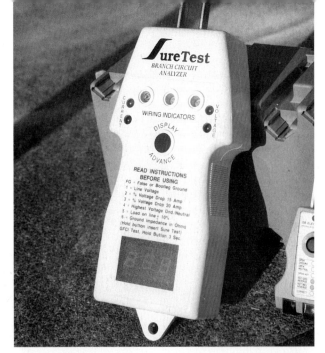

This sophisticated tester is a complete branch-circuit analyzer. It checks for proper wiring, will trip a properly wired GFCI and will indicate the exact trip current to verify if the GFCI is with its current limits.

tripped off. She kept doing this time after time, day after day, click after click, until finally the button broke off and flew across the room.

GFCI Testers

With GFCIs being so popular, many specialty GFCI testers have entered the marketplace in addition to the standard low-cost tester (see the left photo above). A tester places leakage current to the bare wire, grounding system. Some designs will allow the user to vary the current imbalance to see the exact amount of difference current it takes to trip the unit. Others testers place faults at specific levels—for example, .003 amps, .005 amps, .007 amps, .01 amps and .03 amps. However, on the high-end of the line, and the best I've used, is a tester made by SureTest (see the right photo above). Besides giving the entire branch circuit a complete testing makeover, including improper wiring, the tester automatically steps through leakage currents so that you can find the level at which the GFCI will trip.

Remember that you cannot immediately fail a GFCI because it doesn't trip when tested. The unit could

be wired wrong, or the line voltage could be too low to obtain the proper leakage current. In addition, you cannot fail a GFCI if it trips at the wrong current unless you are using a tester that is not voltage-dependent—which is 99% of all testers. A standard tester uses a resistor to find leakage current. Voltage and current are directly proportional: If the voltage goes up and down, so will the leakage current. It's possible that the voltage could go down so low that the leakage current might not trip the GFCI. The SureTest unit, however, uses a thermister to vary the current and is not voltage-dependent. If a GFCI fails with the SureTest, it fails.

Never buy a house—new or old—without first checking all GFCIs with a tester to verify that the units are working properly. I've lost track of the number of defective GFCIs I've found on walk-throughs. If you find the problem before the sell, the seller will have to fix the problem. If you find the problem after the sell, you may have to fix it. And let's hope you don't find out the hard way that the unit is defective.

Chapter 9

RECEPTACLES AND BOXES

RECEPTACLE BOXES

RECEPTACLES

POLARITY

WIRING AND INSTALLATION

You're about to learn all the top-secret information about receptacles, boxes and wiring that contractors rarely discuss. Knowing what to buy and what to avoid, what looks good and what really is good, and what works and what doesn't is information that must be known before receptacle installation begins—not after it's too late.

Receptacles discussed in this chapter will be the standard 120-volt devices seen throughout the home. The receptacles for specific major appliances like stoves and dryers are discussed in Chapter 12. Twist-lock receptacles will not be discussed at all.

RECEPTACLE BOXES

One problem many electricians get into is not thinking ahead and having the right box for the right job. When buying outlet boxes, you should consider many things. Metallic or nonmetallic? Bracket-mount or nail-on (also called captive or integral by the manufacturers)? Deep or shallow? Each has advantages and disadvantages. Not knowing what receptacle box you need leads to many problems, and one of the biggest is using a box that's too small.

Volume Problems

Most receptacle boxes have one common problem: not enough room or volume inside the box to contain the receptacle and the splicing for the incoming and outgoing cables. If the box is too small and doesn't have enough volume, the wires become overcrowded, resulting in broken and shorted wires

Cable Fill

Size of conductor	Volume unit*
14 gauge	2 cu. in.
12 gauge	2.25 cu. in.
10 gauge	2.5 cu. in.

* Volume unit is the space (in cubic inches) that a conductor will occupy in a box.

and possible damage to the box's receptacle-holding threads. To prevent this from happening, the NEC regulates the amount of wires and devices (like a receptacle) that can go into a receptacle box (see NEC Section 370-16).

The volume of space that conductors occupy within a receptacle box is called cable fill. When too many conductors are within the box, it is called a cable-fill violation. Cable fill is calculated in cubic inches (see the chart above).

Using the chart, plan your cable fill. A typical cable, such as 12-2 w/g NM cable, will have three conductors: black, white and ground. All the conductors are added individually, except for the grounding conductors. Regardless of the number of cables entering or leaving the box, the grounding conductors equal one volume unit (2.25 cu. in.) for the total of the box. Add one volume unit total for any cable clamps or support fittings in the box. Add two volume units for the receptacle. Do not add anything for small items like cable connectors and bushings.

Rather than add the volume for individual conductors, I prefer to add the volume for the entire cable because virtually all the receptacle wiring in a house is with NM cables. One 12-2 w/g cable needs 4.5 cu. in. of volume.

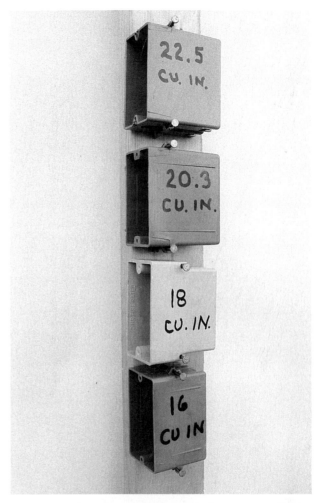

Receptacle boxes come in varying depths. You should choose the deepest allowable to avoid the possibility of a cable-fill violation.

A typical situation is to have one 12-2 w/g cable entering the box, and another exiting. The two cables add up to a volume of 9 cu. in. (4.5 + 4.5). Add 2.25 for all the grounding wires, and 4.5 for the receptacle. If the box is metal, and hold-down clamps are used (add nothing for an NM connector), add another 2.25 cu. in. Therefore, a typical in-and-out situation requires 15.75 cu. in. of volume for plastic, and 18 for metal. Add a third 12-2 w/g cable, and you're up to 20.25 cu. in. for plastic and 22.5 cu. in. for metal, which means you must use a box with significant volume.

Boxes for Shallow Walls

For shallow, furred-out walls, use a box at least 4 in. square and 1¼ in. to 1½ in. deep. Check whether depth is measured from inside or outside box.

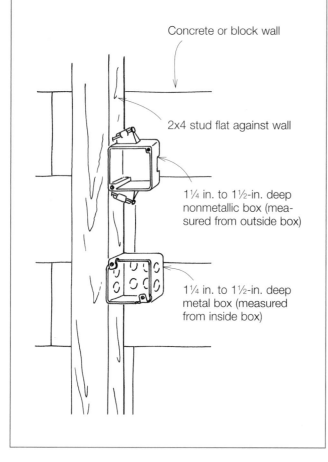

Concrete or block wall

2x4 stud flat against wall

1¼ in. to 1½-in. deep nonmetallic box (measured from outside box)

1¼ in. to 1½-in. deep metal box (measured from inside box)

The deeper, the better A deep box will obviously have more volume than a shallow one, allowing more cables in it. For single-gang boxes (which hold one receptacle or switch), the most common receptacle boxes, a depth of 3¼ in. to 3½ in. is required if you want to have room for three 12-2 w/g cables, along with the receptacle. Such a depth will provide a volume of 22.5 or better, which is preferred (20.3 is barely acceptable, and an 18-cu. in. box can't hold three 12-gauge cables). Boxes with less volume will place you in NEC cable-fill violation almost as soon as you add a second cable.

(Nonmetallic boxes have their volumes stamped inside.) Don't try to downgrade the box size because you assume you will be using 14-gauge wire. You should design your boxes for 12-gauge wire, even if you plan for 14. This way, the worst that can happen is that you will have extra room in the box.

When you can't get depth There will be times when it will be physically impossible to install a deep box. A typical example is when a carpenter furs out a basement wall. To do this it's common for a carpenter to attach the studs flat against the concrete or block wall, making the wall cavity only 1½ in. deep. Since a deep electrical box can't be used unless you knock holes in the wall everywhere you want a receptacle, you must go sideways to get volume. Don't be tempted to use the commonly available metal handy boxes because they seem to fit the depth; they simply don't have the volume for in-and-out cabling.

Assuming that ½-in. drywall is the finished wall, the total allowed depth for the box will be 2 in. A 1½-in. deep, 4-in. square box with a ½-in. cover plate—also called a mud ring—will raise the drywall from the stud by only a small amount, creating only a slight bow after it's installed (you can lessen the bow by using flat-head screws instead of the rounded heads provided by the manufacturer). But beware! Several manufacturers make nonmetallic 4-in. square boxes that are 1⅝ in. deep with ⅛-in. to ¼-in. deep screws and mud rings on top of that. These boxes and mud rings will bow the drywall significantly. Metal boxes, 1½ in. deep, are measured from the inside of the box, which makes them really 1⁹⁄₁₆ in. deep, with another ¹⁄₁₆ in. for the mud ring. This will work fine as long as you don't mind a ⅛-in. bow in the drywall around the box. Or you can use a 4-in. square, 1¼-in. deep box, which will not bow the wall at all. If a square box has a mud ring or faceplate, it must be installed before the finished wall goes up. The finished wall will be cut around the mud ring's raised area, not the box itself.

If you can't find the proper box to fit within the wall cavity, but you can find one that almost fits, you may have to remove some of the solid wall behind the box by scoring it away. Using a circular saw with a masonry blade, make a few shallow cuts in the masonry wall where the box is to be mounted (be sure to wear safety glasses), then swing the saw back and forth to remove the raised edges, and, bingo, a perfect fit. This may sound easy, but if you have to install 30 to 40 boxes, you'll will wish you had spent more time trying to find the proper box.

Metal Boxes

Once the standard of residential construction, metal boxes are used less frequently. The use of nonmetallic boxes is on the rise because they are less expensive, install faster and are nonconductive. However, metal boxes do have three significant advantages over nonmetallic boxes that require electricians to use them: They are strong, have more available designs, and some metal boxes can be stacked, or doubled up. Stacking allows the rectangular switch or receptacle box to increase in size—single, double, triple—by screwing them onto each other (see the top drawing at right). In addition, metal boxes are still needed in areas of the house where the entire box is exposed, such as along a basement or garage block wall, or in a situation where the required design is only available in metal.

Handy, or utility, boxes are the most misused of all the metal boxes. The common rounded-corner 1½-in. deep box only has 10.3 cu. in. of volume available, so a single 12-gauge cable, along with a receptacle, won't fit (11.25 cu. in. would be needed). I don't know of any electrician, including myself, who hasn't installed a 12-gauge cable with receptacle in a handy box, but it is a cable-fill violation. If a handy box needs to be used, pick one that's at least 1⅞ in. deep, or even better, 2⅛ in. deep.

When volume is needed in metal boxes, the 4-in. and 4¹¹⁄₁₆-in. square boxes are the handiest to have around—normal depths are 1½ in. and 2⅛ in. The boxes either have a rounded edge or a sharp edge. The best design (see the bottom drawing at right) allows the outlets to connect directly to the box and utilizes a standard plastic face plate. This box will be the least expensive and fastest to install. A second variety of rounded-edge box (used where depth is a

Stacking Metal Boxes

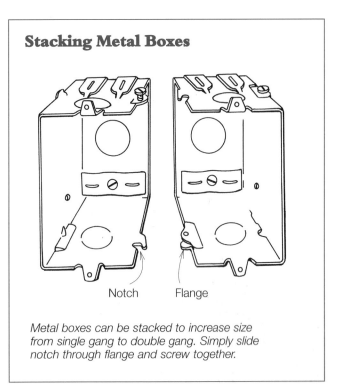

Notch   Flange

Metal boxes can be stacked to increase size from single gang to double gang. Simply slide notch through flange and screw together.

Rounded-Edge Box

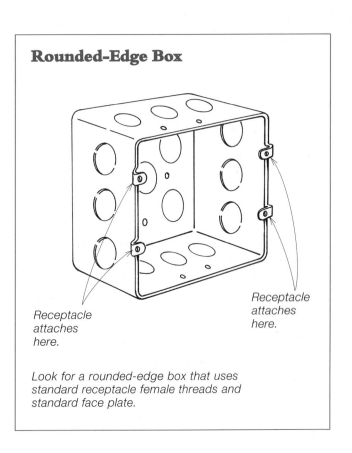

Receptacle attaches here.

Receptacle attaches here.

Look for a rounded-edge box that uses standard receptacle female threads and standard face plate.

Accessories for a Shallow, Rounded-Edge Box

Shallow boxes, both 4 in. and 4¹¹/₁₆-in. square with rounded edges and rounded face plates, are commonly used to house receptacles in unfinished basements and garages, where boxes are not within walls but surface-mounted along block or concrete walls (see drawing on p. 136).

Rounded-edge box

Rounded-edge face plates are available in many designs.

Rounded-edge face plates

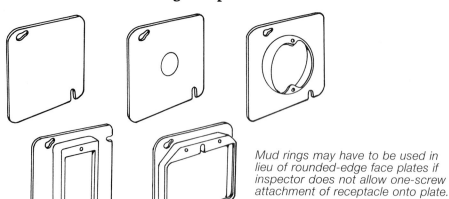

Mud rings may have to be used in lieu of rounded-edge face plates if inspector does not allow one-screw attachment of receptacle onto plate.

Rounded-edge mud rings

problem—see the drawing on p. 136) allows you to attach the receptacle to the face plate, and then the face plate connects to the box (see the drawings on the facing page). The drywall ears will have to be removed for the receptacle to fit within the face plate. However, you must prove that the receptacle is listed for a one-screw attachment. Therefore, you should check with the local inspector before installing this type of box. If your inspector does not allow such an installation, you'll have to use mud rings (raised plate or adapter) to allow a double-screw attachment to the box.

Sharp-edged square boxes are also available, but they don't have much use around a residence. If the work is exposed, a rounded-edge box should be used, which means that the use of sharp-edged boxes is limited to within a wall cavity. And there is no reason to use them there because nonmetallic boxes are the preferred boxes within a wall. In addition, rounded-edge mud rings should not be used on sharp-edged square boxes because the rings leave parts of the box showing (and it looks ugly). When mounting square boxes, be sure to mount the box in such a manner that the face-plate screws will be on the upper right and lower left. Otherwise, the receptacle may wind up horizontal if you don't want to drill out the other two corners of the face plate.

If you're looking to use a metal nail-on box, you'll have trouble finding one with significant volume. And metal nail-on boxes cannot be ganged: You can't make a double or triple by bolting them together. My advice is to forget about using metal nail-on boxes altogether.

For flush-mounting light fixtures on exterior walls, as well as on exposed beams, it's best to use a pancake metal box. A pancake box is a round, 3¼ in. to 4 in. diameter, ½-in. deep box that is mounted on a wall or ceiling, and the light fixture attaches directly to it—there is no face plate on the box. When using such a box, be sure that you do not create a cable-fill violation, which is easy to do because of the low volume of the box. On an exterior, it's best to install the pancake box after the siding is up so that you don't have to worry

Winding a couple of turns of electrical tape around the receptacle sides will insulate the hot and neutral screws from a metal box.

about whether the finished wall will be flush with the box. Simply bring the wire through the sheathing on the exterior wall. The siding installers will drill a hole in the siding and push the wire through. Then you can install the pancake box around the wire (be sure the box is attached to something secure) and install the fixture (for more on pancake boxes, see p. 184).

Metal boxes are conductive There is always the risk of shorting the hot receptacle screws against the sides of a metal box, or having the wires cut into the metal as the receptacle is being pressed back into the box. This is extremely common when a GFCI replaces a standard receptacle in a metal box. Many electricians cross their fingers and look the other way as they shove the GFCI into its too-small metal box. One of my helpers blew out an entire circuit doing this. To prevent the screws from shorting onto the sides of the metal as the receptacle is being pushed in, I always place a couple of wraps of electrical tape (3M Super 88) around the receptacle, covering the hot and neutral screws (see the photo above). An additional trick-of-the-trade is to place a thin insulator board (commonly found in electronics stores) in the back of the metal box to

Nonmetallic Boxes

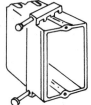

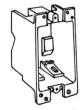

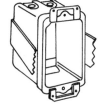

**Single-gang
nail-on boxes**

**Single-gang
cut-in boxes**

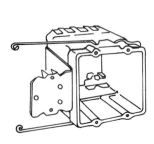

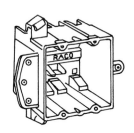

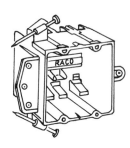

**Double-gang
nail-on boxes**

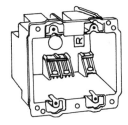

**Double-gang
cut-in box**

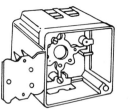

**4-in. square boxes
with single- and double-gang
mud rings**

When securing a nail-on plastic box to a stud, make the last few hammer hits light, or the box will bow.

The front edge of the box should be even with the front edge of the stud, not jutting out as it's shown here.

Be careful when driving the nails for the box into the stud, or you could destroy a box. This one was broken on the last hammer hit.

keep the wires from being cut on any sharp metal edges. If such a board isn't available, stick some electrical tape into the back of the box.

Nonmetallic Boxes

Nonmetallic boxes are the most common receptacle boxes used in residential construction (see the drawing on the facing page). They are both inexpensive and nonconductive. Although material to make the boxes will vary from manufacturer to manufacturer, the most common materials are flexible PVC plastic, fiberglass and Thermoset. These boxes are normally used throughout the house where wires are kept within walls. However, the common rectangular box, such as the integral-nail box, should not be used if the entire box has to be surface-mounted in a habitable area. In addition, the common rectangular, nonmetallic box is *not* to be used for hanging lights. To hang lights the box must be listed for that purpose. Receptacle boxes differ from switch boxes in that it is rare to have more

than a double gang (two receptacles in a box). Switches will commonly use three- and four-gang boxes for all the switching required at entryways (for more on switches and boxes, see Chapter 10).

The most used is a single-gang, nail-on box. For this type of box, simply nail the box to the stud and leave just enough sticking out from the stud to be flush with the finished wall. Three challenges occur here (see the photos above): Getting the distance exactly right out from the stud, keeping the box parallel to the stud, and keeping the box from being distorted or broken as it is nailed onto the stud. To prevent the first two, take your time as you install the box. Some boxes will even have raised edges or ribs to help you judge the distance out from the stud (see the drawing on p. 142). In addition, you can use a piece of ½-in plywood as a template to help get the distance right. Temporarily attach the plywood to the stud and align the edge of the box with the plywood. To prevent breaking the box as you nail it

Nail-on Box with Ribs

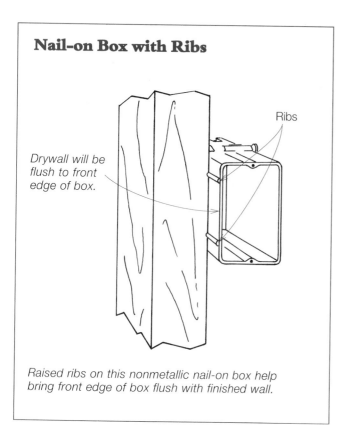

Drywall will be flush to front edge of box.

Ribs

Raised ribs on this nonmetallic nail-on box help bring front edge of box flush with finished wall.

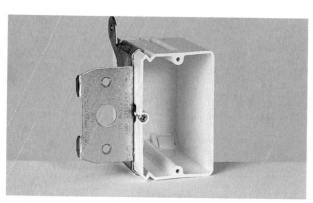

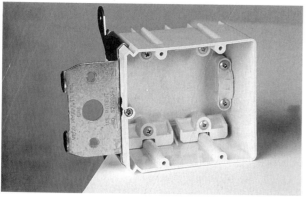

in, simply tap the nails—don't drive them hard—as the nail head comes close to the box. The more accurate you get the box mounted on the stud, the less trouble you'll have making the receptacle fit up against the finished wall. If you don't want to be bothered by constantly solving these problems, install the slightly more expensive side-mount box, which has a bracket mounted to its side that nails to the front of the stud, placing the box at a prescribed distance (normally ⅜ in.) out from the stud and keeping its face parallel to the stud. This box will extend beyond a thin finished wall, however.

A new type of nonmetallic box allows you to adjust the box in or out with the turn of a screw—after the box has been installed. Made by Veco Products, Inc., the box can slide backward or forward to align with the finished wall. Although it sounds too good to be true, these boxes work, and they are perfect for the times when you don't know what the finished wall will be—tile, drywall or wallpaper. Currently, these boxes are available in single- and double-gang designs (see the photos below left).

Cut-in Boxes

For installing receptacle boxes in existing hollow walls, special boxes, called old-work or cut-in boxes, are used (see the top drawings on the facing page). These boxes do not mount to studs; instead, they attach directly to the finished wall, using plastic ears in front and supports (either swivel ears or wings, also called spring ears or clamps) in the back. The plastic ears hold to the drywall surface, and the supports in back sandwich the drywall to hold the box in place. In addition, you can make any box with drywall ears work like a cut-in box, as long as you have an adapter to do so (see the bottom drawing on the facing page). And remember, if you're installing a cut-in box in a wall with a receptacle directly on the other side, offset the new box a bit so that the boxes are not back to back—

These nonmetallic boxes from Veco Products, Inc., can be adjusted in or out with the turn of a screw. Each box can be moved flush with the finished wall after it is nailed in. (Photos courtesy of Veco Products, Inc.)

Cut-in Receptacle Boxes

Cut-in boxes are used mostly in remodeling. They attach to face of finished wall through drywall ears on front. Support mechanism on back sandwiches box to wall.

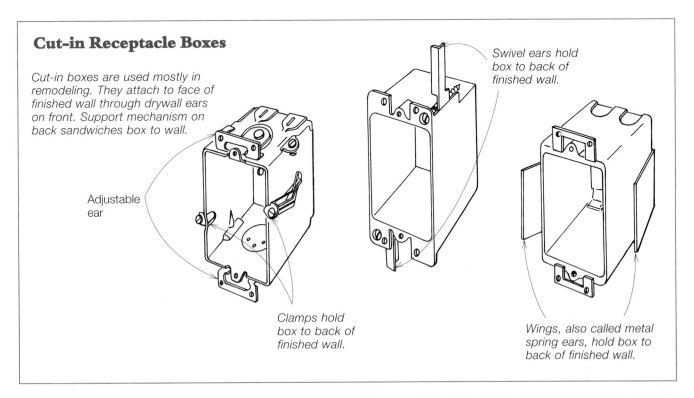

Swivel ears hold box to back of finished wall.

Adjustable ear

Clamps hold box to back of finished wall.

Wings, also called metal spring ears, hold box to back of finished wall.

---

usually there's not enough room in the wall cavity to do this (see p. 109). This will leave you with plenty of room to work in the wall cavity.

Cut-in boxes allow for few errors (see the drawing on p. 144). The attachment system on the cut-in box will only work if the hole is small enough for the box's plaster ears to press onto the finished wall front. If the electrician is sloppy and cuts the box hole too big, the box will simply fall through the hole. A good electrician will cut the hole exactly to fit the box. Some manufacturers furnish templates to aid in cutting out the hole. However, with other manufacturers, you're on your own.

No matter how good an electrician you are, each one of us will eventually cut the hole too big for the box. The only question is how to fix it. A trick-of-the-trade is to cut a thin cedar shim (the kind used to shim doors and windows) 2 in. to 3 in. wider than the hole. Put some glue on the shim, slip it into the hole and pull it back against the inside of the finished wall. Then cut a piece of finished wall the

Cut-in Adapter

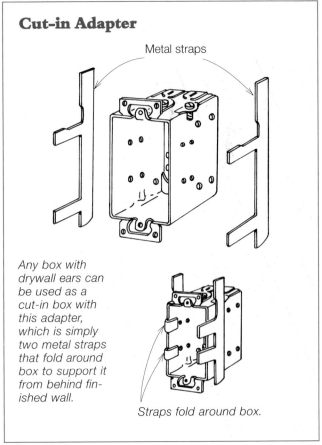

Metal straps

Any box with drywall ears can be used as a cut-in box with this adapter, which is simply two metal straps that fold around box to support it from behind finished wall.

Straps fold around box.

Installing a Cut-in Box

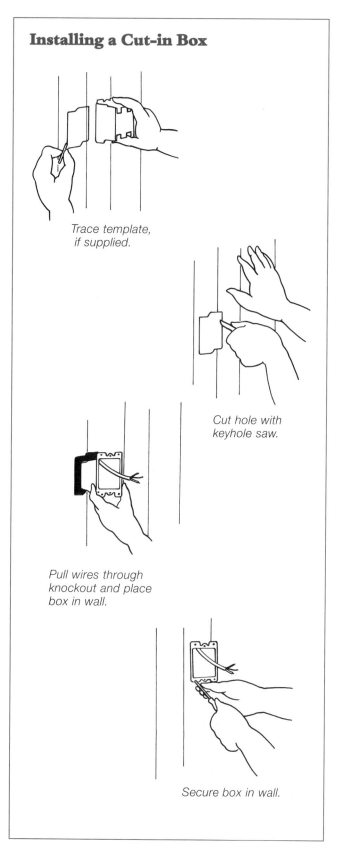

Trace template, if supplied.

Cut hole with keyhole saw.

Pull wires through knockout and place box in wall.

Secure box in wall.

exact size needed to make the opening the correct size and glue it on top of the stick.

The problem with all cut-in single-gang boxes is that they don't have much volume, with 10 cu. in. to 18 cu. in. being common. This severely limits the number of cables the box can hold—normally only to one or two. However, the 10-cu. in. and 12-cu. in. boxes are ideal for narrow wall cavities because they are extremely shallow.

If you're feeling lucky and want to be a bit adventuresome, you can take virtually any box, even those without drywall ears, and use it as a cut-in box. Drill a small hole in the finished wall adjacent to a stud and stick a wire in to get the stud's exact location. Once located, open a small hole in the wall so that you can see the stud. Place the box adjacent to the stud and trace around the box. Next, cut a hole and slip in the box (don't forget the fish wire to slip the cable into the box) and nail or screw it to the stud. I prefer using drywall nails because their thin, flat heads don't stick into the box very far, and they have raised edges on the shank so that they can't easily back out. I normally predrill with a small bit so that I won't crack the box (predrilling also allows the nail to go through the box and into the wood easier). Sometimes I put some adhesive between the box and the stud for added insurance. Whether you use nail or screws, the heads must be as flush to the box inside as possible so that they don't hit the receptacle screws. The advantage of following this route of adventure is that you can pick your boxes rather than having to use the common, low-volume, premade, cut-in boxes.

Weatherproof Boxes and Covers

Made from cast metal, weatherproof boxes are used outside the house when the entire receptacle box is to be exposed. Because of aesthetics, new construction locates most receptacles boxes for outside outlets inside the exterior walls. This allows a less-expensive standard box to be used in the wall protected from the elements by a weatherproof cover. When adding an outlet to a finished outside wall, a weatherproof box is normally attached to the outside siding. Even though the box is weatherproof, the receptacle still needs to be protected with a weatherproof cover (see the drawing at right). Though the standard single-gang cast-metal box is the most popular, a great variety of boxes is available. Nonmetallic weatherproof boxes are normally used with plastic glueable conduit. The conduit is easy to install, comes in all common diameters and can be used with standard weatherproof covers.

There are two types of weatherproof covers: weatherproof while the cover is closed and weatherproof while the receptacle is being used. For intermittent use of an outside receptacle while the sun shines, a standard weatherproof cover is acceptable. For continuous use—for Christmas lights, fence charges and low-voltage transformers—a weatherproof while-in-use cover must be used (see the photo below right).

RECEPTACLES

In residential construction, cheap, low-grade receptacles typically get used for both light- *and* heavy-duty applications. Because of the poor quality of low-grade receptacles, I don't use them for even light-duty applications. They can break apart, cause open circuits, or even worse, cause a high-resistance loose connection that results in a fire. During the writing of this chapter, I was called out on a job where the customer had a burned-out receptacle. He first smelled something burning (the insulation on the wires) and then pulled the appliance away from the wall to see the receptacle glowing cherry red. If he was not home, his house probably would have burned to the ground. His problem was a cheap receptacle and the way it was wired. Instead of worrying about color and style, homeowners and electricians should worry more about quality.

Grades

There are several different grades of receptacles. The lowest grade, the type you normally get, may be fine for light-duty, rarely used situations such as in a bedroom or hallway. But for more heavy-duty applications, such as in a kitchen, garage or shop, a heavy-duty receptacle should be used. Residential, commercial, hard-use, specification, federal spec

Outdoor Outlets Must be Covered

Exterior siding

Standard nail-on box within wall cavity

Cast-metal weatherproof box outside exterior siding

Both boxes must be protected with weatherproof covers, and GFCI-protection is required.

This weatherproof while-in-use cover by TayMac allows an appliance to be plugged in while the cover is closed.

grade, industrial and hospital grade are just some of the names manufacturers give their receptacles to differentiate their use and construction. You don't have to take what you are offered by a salesperson: Ask for what you want.

Contractor grade Low-bid jobs normally reflect the lowest-quality receptacles, commonly called contractor or residential grade. This receptacle has a bad habit of breaking apart as wire is fastened to it, of having the plastic around the ground pin break off and of having the female terminals strip out as the screws tighten down on the wires. Some even have screws that are too small to hold the wire tight. Contractor-grade receptacles are made from very thin and sparse metal. Quality is so poor that many times I've had this type break apart in my hands before I could get it into the wall!

Commercial grade These receptacles, also called spec grade, are high-quality receptacles on a budget. Though some of these models go for high prices, others are more moderately priced. For a small increase in price, about 50 cents to $1 more than the bottom of the line, you can get a hard-to-break nylon face, a wrap-around yoke for receptacle support, a heavy-duty body, a self-grounding clip and heavier metal on the inside. Most manufacturers make these alternatives available; you just have to ask for them. An indication of higher quality is a wrap-around yoke on the back of the receptacle— manufacturers don't normally put this on contractor grades. Moderately priced spec- or commercial-grade receptacles should be the lowest-grade receptacles that you put in your house. Some manufacturers make a spectrum of colors: yellow, red, ivory, white, gray, brown and black. Others will only have these available in ivory and brown.

The ultimate receptacle? On the high end of the quality scale is the Hubbell 5262. This receptacle is the reference for quality. It is so good that when Hubbell invented it in 1950, many companies copied the model number so that they could get in on some of the market when architects simply said to order a 5262 (you can't patent a model number). Hubbell

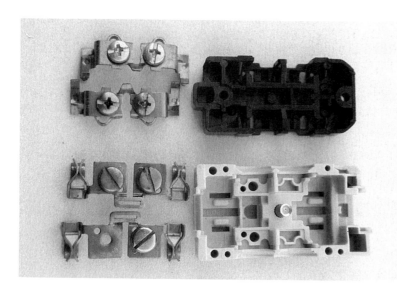

In the top of the photo are some parts of a light-duty receptacle. In the bottom of the photo are parts of a heavy-duty model, which has a heavier plastic body, a self-grounding clip and heavier metal on the screw terminals.

Commercial-grade receptacles offer more quality for the money. This one has a wrap-around yoke, which is the solid metal strap on the back, and accepts four hots and four neutrals. The wires are inserted from the back, and screw pressure on a metal bar keeps them tight.

now tells customers that if you want a true 5262, you must specify a Hubbell 5262 (see the left photo below). Costing around $15 to $20, which is the price of some decorator receptacles, you would infinitely be better off spending your money on quality rather than beauty. What you get is a receptacle that can take eight wires (four on each side) that fit under a metal plate and tighten down via a screw. It is flared on each side to prevent the side screws from shorting against a metal box, as well as nonbreakable, very-heavy-duty everything. If the price of the 5262 is totally out of the budget, you might consider placing heavy-duty receptacles in the heaviest-use areas of the house, and light-duty receptacles in light-duty areas. More price-competitive Hubbell receptacles are the 5252 and 5242. I've used both, and they are excellent receptacles. All three receptacles come in colors of brown, ivory, red, gray, white and black.

Heavy-gauge wire requires heavy-duty receptacle

Most receptacles will take 14- and 12-gauge wire with no problem—but beware of 10 gauge. Many receptacles will break apart, or the screws will strip out when 10-gauge wire is attached. In addition, even though a 10-gauge wire may safely conduct 30 amps, a standard 15- or 20-amp receptacle cannot. In my opinion, 10-gauge wire should only be installed on commercial-grade receptacles or better. In addition, a fancy grade rating is immaterial if the receptacle face is still made from cheap, brittle plastic. When I built my house, I installed commercial-grade receptacles in my garage. Later, several of them broke apart as I wiggled my heavy-duty extension cord plug up and down to get it out. As each broke apart, I replaced it with a Hubbell 5252 or 5242. Lesson learned: Be sure the face is made from a nylon or other impact-resistant plastic.

Duplex or Single

A duplex receptacle has two outlets to plug into. A single receptacle has only one (see the right photo below). There are many instances where you may want or even need to use a single receptacle, as opposed to a duplex. For example, you may want to put your freezer or refrigerator in an unfinished

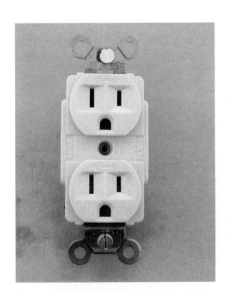

The ultimate receptacle is the Hubbell 5262. The side flares to keep receptacle screws away from the box.

This is a single receptacle rated at 15 amps.

basement, where receptacles are supposed to be protected by GFCIs. But you'd rather not have these appliances hooked up to a GFCI receptacle because of possible false tripping. A cord-and-plug appliance like a freezer does not have to use a GFCI receptacle in an unfinished basement if it's connected to a *single* receptacle, not a duplex. Two appliances hooked to a duplex receptacle is also acceptable.

A duplex receptacle has a tab on both hot and neutral sides. When half the receptacle is to be switched, and the other half is to be permanently

hot, break off the tab to isolate the top receptacle plug-in from the bottom (see the top photo at left). Insert a screwdriver into the slotted tab and wiggle it back and forth until the tab breaks off. The tab is normally broken only on the hot side, allowing the neutral to be shared. If a single outlet is required in a certain situation (as in a refrigerator in a basement), but the outlet is unavailable, some inspectors will allow you to break the tabs off both sides of a duplex receptacle to deaden the unused half. This is sometimes allowed because the intent of the code is being followed as opposed to the written word.

Ganged Receptacles, Raceways and Outlet Strips

Kitchen countertops, stereo centers, computer workstations, garages and workbenches all require a large number of receptacle outlets in what may be a small space. Double- and triple-gang receptacles can be wired together, but it's very time consuming and ugly when you finish. Do-it-yourselfers find it especially hard to get all the wiring for multiple receptacles into the box. Some folks may strip the screws that hold the receptacles in the box in an attempt to get all the wires inside it. And if you decide to use a triple-gang receptacle, the cover may be hard to find.

Breaking the slotted tab on a duplex receptacle isolates the top and bottom outlets.

A premade quad receptacle provides four heavy-duty plug-in connections in both 15-amp and 20-amp versions. (Photo courtesy of Bryant)

Attaching a Quad Receptacle to a Single-Gang Box

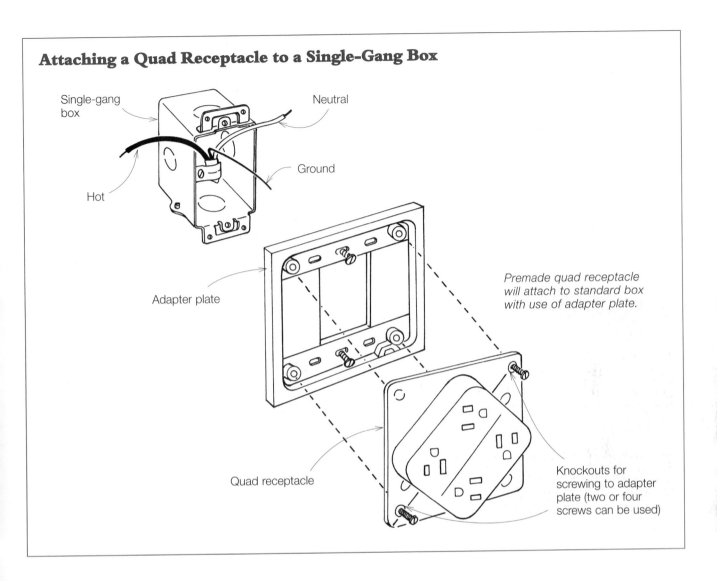

Single-gang box

Neutral

Ground

Hot

Adapter plate

Premade quad receptacle will attach to standard box with use of adapter plate.

Quad receptacle

Knockouts for screwing to adapter plate (two or four screws can be used)

An alternate solution to ganging receptacles is to buy a premade quad receptacle (see the bottom photos on the facing page). This type of receptacle is available from several manufacturers. All the wiring to each receptacle in the quad is already done; you just have to wire the hot, neutral and ground from the main panel. A typical quad receptacle mounts to a 4-in. square or octagonal box. Or it can attach to an adapter plate, which in turn, can connect to a standard single- or double-gang box, so a quad can be installed even after the walls are up (see the drawing above). It provides four heavy-duty plug-in connections in both 15- and 20-amp versions. It installs much faster and is much more attractive

than a standard multiple-gang hookup. Some units are even equipped with LED lights to indicate when power is on. Quad receptacles are available in black, brown, red, blue, white, orange, ivory and gray.

If power is required over a long horizontal area, it's best to install a prewired raceway, which places a receptacle every few inches along its track (see the left photo on p. 150). Coming in various lengths with 3 ft., 4 ft. and 5 ft. being the most common, raceways can be coupled together and are perfect for mounting above kitchen countertops and shop workbenches. Simply bring the rough-in wire out of the wall at the exact location where the raceway is to

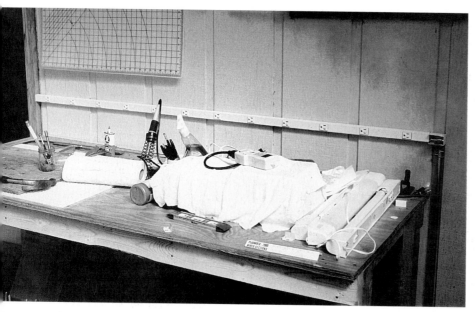

A prewired electrical raceway is screwed into a mounting frame, which is screwed into the wall. In this case, the raceway is located above a workbench. The splice can be made at either end, but in this case it's on the right.

A 20-amp receptacle has a horizontal slot on the neutral terminal to accept a 20-amp plug. It will also accept a 15-amp plug.

be mounted. Splicing will be done in the raceway or in a box.

Plug-in outlet strips, normally seen behind computer desks, are available for any multiple-outlet purpose and provide approximately four to six outlets per strip. Use these instead of cheap adapters or any type of push-in, receptacle-mount multiple tap. I was on a service call once where a push-in multiple tap started smoking heavily—I pulled it loose with my long-nosed pliers before it caught fire.

Current Ratings

Receptacles are rated for the amount of current allowed to flow through them. Standard duplex receptacles used throughout a residence are rated for 15 amps—even those that are connected into a standard 20-amp branch circuit. A 20-amp receptacle looks different than a 15-amp receptacle. The 20-amp receptacle has an additional slot going sideways off the neutral slot (the wide one) to allow a 20-amp

plug with a sideways prong to fit into it (see the right photo above). Both a 15-amp and 20-amp plug will fit into a 20-amp receptacle, but a 20-amp plug will not fit into a 15-amp receptacle. The receptacle is designed this way so that you don't put a 20-amp load on a 15-amp circuit, which will cause the wires to overheat and the breaker to kick. You are required to use the more-expensive 20-amp receptacle when there is only one receptacle in the entire 12-gauge, 20-amp circuit, and that receptacle is a single receptacle.

A significant safety problem occurs when you purchase an appliance without thinking about how much current it pulls. You get the appliance home and find the appliance's 20-amp plug doesn't fit your 15-amp outlets. The appliance salesman may have even told you that all you have to do is change the outlet from a 15 amp to a 20 amp so that the plug will fit. Don't believe him! Replacing the receptacle is *not* enough. A new circuit with 12-gauge cable will have to be installed from the 20-amp appliance receptacle all the way back to the main panel. And

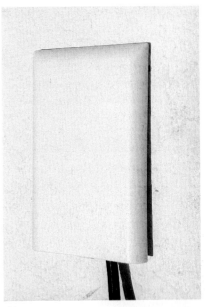

A decorator receptacle is nice to look at, but expect to pay two to four times the price of a standard unit (even though it may be a simple, residential-grade receptacle).

An alternative to buying a residential-grade decorator receptacle is a decorative face plate made by TayMac.

This cover, made by TayMac, hides the receptacles and prevents debris, water and children from entering readily.

let's hope there is still room in the panel for another circuit. A simple rule to remember: Think ahead!

Decorator Models

Decorator receptacles, which have a flat front for the entire face of the receptacle, are becoming more popular (see the left photo above). Gray, white, ivory, red, almond and black are the most common decorator colors. Expect to pay premium price for a decorator receptacle, even though it may only be residential grade. Decorator styles are available in commercial grade just like standard receptacles, so if you want to spend money on fancy receptacles, I suggest you spend the extra money and buy quality. But think ahead: Most hardware stores will stock only residential grade, meaning that you'll have to special-order the higher-grade receptacles. If the price is too high, use a moderately priced, standard commercial grade as opposed to decorator-style commercial grade.

There are alternatives to decorator receptacles. One is to use a standard receptacle and a hide it with a decorative one-piece face plate (see the middle photo above). The reason these have been put on the market is that some homeowners have complained that even with decorative outlets, the break between the receptacle and cover is still noticeable. Another alternative is covering the receptacle with a flat flush cover plate (see the right photo above). When power is needed, open the cover, plug in the appliance, then close the cover. The appliance cord extends out from the bottom of the cover. The cover not only makes the area more attractive, but it also keeps liquids and other materials from getting into the receptacle (as well as fingers of little children). The covers come in single- and double-gang sizes and in several different colors.

Specialty Models

For special situations, you can buy specialty receptacles. For example, a tamper-resistant receptacle will keep children from inserting objects into the receptacle and receiving a shock. This type incorporates a unique shutter or swivel mechanism designed so that a child cannot come into contact with an energized wire unless he or she inserts something into both blade slots at the same time or rotates a cover. A second specialty design will not allow the insertion of an ungrounded plug. Illuminated, surge protected and corrosion-resistant receptacles are also available. An extremely attractive tamper-resistant design keeps the outlet out of harms way by placing it within the wall cavity. Also, the receptacle is under a latched cover, so a child's

fingers cannot get into the receptacle slots, nor can she handle plugs that are halfway in the receptacle with the blades exposed.

Aluminum wiring connecting into receptacles presents special problems. Because aluminum has different characteristics than copper, no receptacle may accept aluminum wiring unless it is designated on the receptacle as AL/CU. This means the receptacle is UL-approved for copper or aluminum wiring. Leviton and Eagle manufacture such

Wire Connectors

Wire nut is the name most electricians give to the little plastic wire connectors that we use for most of our small-diameter (10-gauge and smaller) copper wire splicing. These connectors make a fast, easy and inexpensive way to splice. Splices also can be crimped together, with an insulating sleeve slipped over the crimp. However, such a splice will not allow you to add or take away a wire without destroying the splice and redoing it.

Most connectors are made with a thermoplastic shell over an expanding square-wire spring. The spring cuts slightly into the wire's surface and supplies the inward pressure to make a tight connection. Shells can be round or with wings. I prefer the latter because it gives a more secure place for my fingers to grip the connector to tighten it onto the bunched wires.

Types of connectors
Ideal originated the original Wire Nut over 60 years ago (Ideal owns the trademark). Most manufacturers make winged and round designs. The winged designs normally come in the standard four colors (each

Typical wire nut

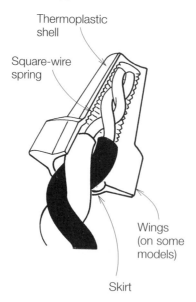

Thermoplastic shell

Square-wire spring

Wings (on some models)

Skirt

color for different wire combinations) that electricians are used to: yellow, red and blue—and green for grounding. The round designs are a little less expensive, and they come in the same four common colors as their winged cousins (they also come in two or three smaller sizes as well). I normally keep one box of each smaller size just in case I need to splice to fixtures. In general, yellow wire nuts can splice three 12-gauge wires or four 14-gauge wires; the red can splice three 10-gauge, four 12-gauge or five 14-gauge wires; the blue can splice four 10-gauge, five 12-gauge or six 14-gauge wires.

Recently, Ideal came out with a design called Twister in two models (tan and gray), which can use a standard nut driver to tighten the wire nut. A significant Ideal development in the Twister line is the UL-listed connector (purple) that can splice aluminum to copper. This connector comes with an internal compound that cuts the

receptacles. If you don't have a receptacle that accepts aluminum wire, you'll have to splice a 6-in. 12-gauge copper pigtail of the appropriate insulation color onto the aluminum wire in the receptacle box (be careful not to put a severe bend on the aluminum wire because it breaks easily). Then terminate the copper wire onto the receptacle screws. Ideal is marketing a new style of wire nut that is listed for copper-to-aluminum splicing—however, it is not listed for aluminum-to-aluminum splicing at this time (for more on splicing and wire nuts, see below).

Special combination receptacles, having different voltage combinations on each half, are available for many specialty situations. One common situation would be roughing in a 240-volt receptacle for a window air conditioner. You also need to have a 120-volt receptacle in this same location to satisfy code requirements. Rather than have two receptacle boxes side by side, you can put both receptacles in the same box.

surface oxide on the aluminum wire for a good corrosion-free connection. This connector, however, is not listed for aluminum-to-aluminum connections.

3M makes a connector with a long, flexible skirt. The skirt provides better protection against bare wires sticking out of the connector and bends with the wire. This allows for a faster turning radius when getting the wire from point to point and better wire placement within the box. All 3M's connectors have wings for added turning leverage. Personally, I like the new connectors, but only time will tell if electricians will take the new design to heart.

King found a niche in the connector market by supplying wire connectors listed for wet locations. Moisture is the enemy of all electrical splices and must be kept out of the connection at all costs. One of the lessons electricians learn along the way

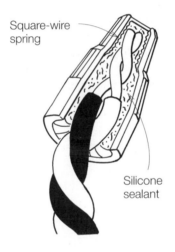

Wire nut for wet locations

Square-wire spring

Silicone sealant

Silicone sealant protects against sparks, water and corrosion. Sealant will not leak or harden.

is that watertight boxes are not watertight. Many times I've had to drill weep holes into the bottom of alleged watertight boxes just to let the water out

(water vapor enters, condenses on the electrical wiring and then pools in the bottom of the box). Previously, I had been squirting silicone into wire connectors to keep the splices dry. King makes connectors already filled with silicone. All electricians should keep these indispensable items in stock for splicing in basements, well casings, outbuildings, crawlspaces and outside lights. The light-blue model (#6) is listed for direct burial. However, the wire itself, not just the connector, must also be listed for direct burial. UF, the most common buried wire or cable used for buried service, is listed for such only when the sheath is intact. A UF buried splice must be inside a watertight box.

Making a good splice
Pull all the wires that are to be spliced together out and to one side of the box; then cut them the same length. Make sure the wires come straight out and that they do not intertwine or loop in

POLARITY

Plugs and receptacles are polarized when they fit together only one way. The neutral blade of the plug (the wide one) must fit into the neutral (wide) slot of the receptacle, and the hot (narrow) blade of the plug must fit into the hot (narrow) slot of the receptacle. Making the neutral blade and slot wider than the hot blade and slot ensures that the hot and neutral wires cannot be reversed for an appliance whose plug doesn't have a grounding pin.

Plugs and receptacles with grounding pins are automatically polarized because of the grounding pin. When the plug is inserted into the receptacle so that the grounding pin of the plug aligns with the grounding slot of the receptacle, the hot and neutral blades will slide into their corresponding receptacle slots. Some motorized appliances, such as fans and double-insulated drills, normally don't need polarization because a motor doesn't care which wire is which; plus the internal wiring is insulated from

Wire Connectors (*continued*)

and out with other wires. Don't make the wires to be the same length by forcing some of the wire back into the box—the extra wire length takes up too much space in the box. Once cut, strip the insulation off the wire ends. The amount of insulation to be stripped will depend on the size of connector you're using. The stripped ends need to have the entire bare wire inside the upper body of the connector, and the wire's insulation needs to be intact within the skirt. Never have bare wire sticking out from the skirt of the connector.

Make a habit of twisting the wires together with your electrician's pliers (in a clockwise motion) before twisting on the connector. You will know when to stop twisting the wire nut when you feel significant resistance. If you don't feel much resistance, the connector is too big—use a smaller one. If the wire nut won't fit, use a larger one.

If the splice is not tight, a fire can occur due to arcing. Though a standard shell won't support fire and is rated up to 105°C/221°F, a loose connection will destroy any connector to the point where the arcing wire's flash can start a fire. I've seen this happen many times.

Once the splice is properly made, fold the wires neatly back into the box. Have a system:

Don't just haphazardly push the wires back into the box. I always fold the grounds back first to get them out of the way. Then I push the hot wires into one side of the box, the neutrals to the other. I try to keep the splices to the outer perimeter of the box so that they won't be exactly behind the switch or receptacle.

Splicing with a wire nut

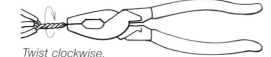

1. Cut all wires the same length.

2. Strip off enough insulation that no bare wire sticks out of the connector.

3. Twist wires together clockwise, using broad-nosed electrician's pliers. Then insert wire nut over wires and twist clockwise.

Twist clockwise.

Polarity of Receptacles and Plugs

In an electrical circuit, neutral and hot should be kept separate, which is called polarization. For a plug to be properly polarized, its neutral and hot wires must plug into the neutral and hot slots, respectively, of the receptacle.

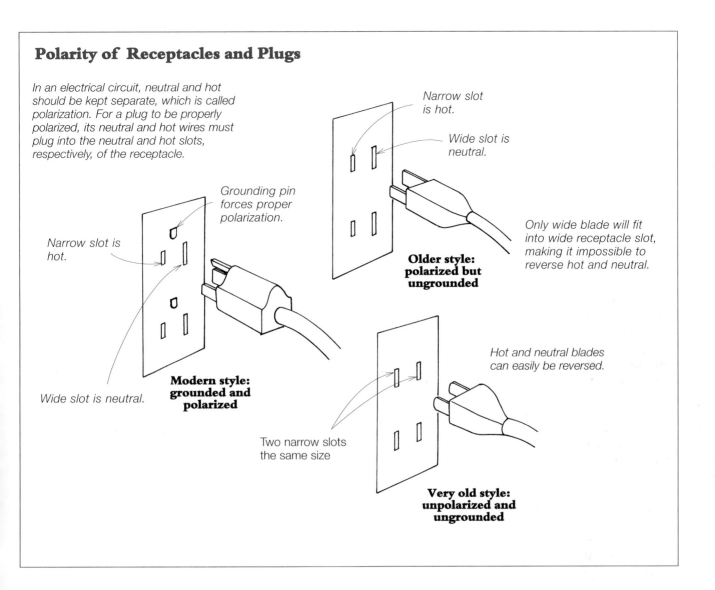

Narrow slot is hot.

Wide slot is neutral.

Grounding pin forces proper polarization.

Narrow slot is hot.

Only wide blade will fit into wide receptacle slot, making it impossible to reverse hot and neutral.

Older style: polarized but ungrounded

Wide slot is neutral.

Modern style: grounded and polarized

Hot and neutral blades can easily be reversed.

Two narrow slots the same size

Very old style: unpolarized and ungrounded

---

all metal parts. Such an appliance has an ungrounded, unpolarized plug, with both blades of the same width.

Safety Precautions

Some disasters are just waiting to happen. Sitting on my desk is a table lamp connected into the receptacle via an old-fashioned two-conductor cord. The plug has two narrow, unpolarized blades, so it can be inserted either way into the receptacle, and the lamp will work just fine. However, for safety reasons, the center contact of the lamp should be connected to the hot wire of the branch circuit (it's

also a code requirement)—and the screw base of the lamp to the white neutral. But because there is no polarization on the plug, you have a 50/50 chance of getting it right—or wrong. If the bulb is out of the lamp, and the screw base is hot because of incorrect polarization, my inquisitive grandchild could easily reach in and touch the inside of the screw base and be electrocuted. (A similar accident recently killed a child in Virginia.)

To be sure this doesn't happen to your children, do what I have done: 1. Always leave a bulb in a lamp that is still plugged into the wall. 2. Check the

Checking Polarity of an Unpolarized Plug

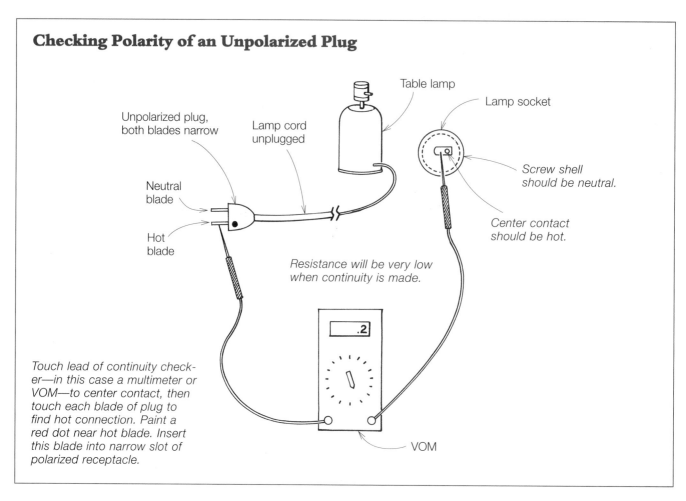

Table lamp

Lamp socket

Unpolarized plug,
both blades narrow

Lamp cord
unplugged

Screw shell
should be neutral.

Neutral
blade

Hot
blade

Center contact
should be hot.

*Resistance will be very low
when continuity is made.*

.2

*Touch lead of continuity check-
er—in this case a multimeter or
VOM—to center contact, then
touch each blade of plug to
find hot connection. Paint a
red dot near hot blade. Insert
this blade into narrow slot of
polarized receptacle.*

VOM

Receptacle Orientation for Immediate–Turn Plug

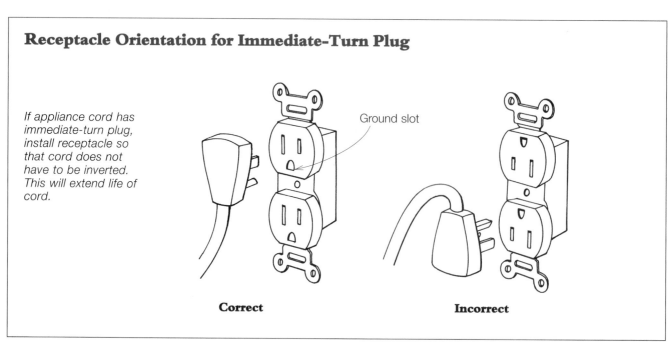

*If appliance cord has
immediate-turn plug,
install receptacle so
that cord does not
have to be inverted.
This will extend life of
cord.*

Ground slot

Correct

Incorrect

continuity between the center contact of the light's screw-in base and the plug blades to determine which blade is connected to the center contact (see the top drawing on the facing page). Once determined, put a dot of paint on that plug's side and then align the paint dot with the small slot on the receptacle. You now have correct polarization: Hot to center contact, and neutral to screw base.

Old-House Polarization

Old houses with unpolarized receptacles present a unique problem. All electricians know there is no simple way to tell hot from neutral when the wiring is so faded that you can't tell black from white. A good trick of the trade will tell you what's what. Run a wire from the panel neutral to one lead of a VOM, and the other VOM lead to one of the receptacle slots. The neutral slot should read 0 volts, and the hot slot should read 120 to 125 volts. If the system is in conduit, the conduit can be a ground reference for the test. Without conduit, such old systems are, of course, ungrounded, which brings a second danger into the formula. If budget permits, the wiring of such old houses should be upgraded for safety.

WIRING AND INSTALLATION

The wiring and installation of receptacles is one of the most error-prone areas in residential wiring. Receptacles are installed upside down, polarities are reversed, receptacles have the wires attached to them in an unsafe manner, and the wiring is installed to the receptacles in a design akin to cheap Christmas-tree lights.

People always ask me, "Does the grounding terminal go up or down?" and "Is there a bottom or top to a receptacle?" There is no "official" right or wrong way to orient the receptacle—the NEC doesn't specify—but one can decide by using logic and common sense. For example, clothes washers, refrigerators and window air conditioners typically have cords with immediate-turn plugs. For appliances with this type of plug, orient the receptacle so that the plug can insert into the receptacle and not have to loop over itself (see the bottom drawing on the facing page).

Positioning Receptacles

With grounding slot on top, falling metal object, such as picture frame, will bounce off a partially exposed plug with no harm done. With grounding slot on bottom, object could short across hot and neutral blades of plug.

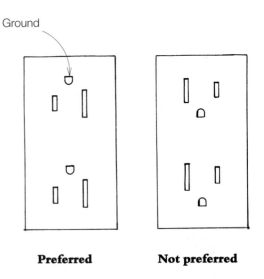

Preferred   **Not preferred**

Receptacle installed horizontally should have neutral (wide slot) on top. A falling metal object will make contact with neutral blade of partially exposed plug, not hot.

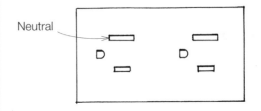

Preferred

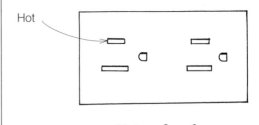

Not preferred

Otherwise, the downward pull will tend to pull the plug out of the receptacle.

Other than for immediate-turn plugs, I prefer to put the grounding slot on top. Reason being, if a plug was partially pulled out of the receptacle, exposing the hot and neutral blades of the plug, and something metal fell on it, a direct short would not occur because the grounding pin would deflect the item away from the hot terminals (see the drawing on p. 157).

Receptacles can also be positioned horizontally. When this situation occurs, place the grounding slot to the left, and the wide neutral slot on top. The logic of this is if a plug has its blades partially exposed and something metal falls onto it, like a picture frame, it will hit the grounded neutral blade as opposed to the hot.

Identifying Screw Terminals

A receptacle has three locations to accept wires. 1. A hot side with two gold screws internally connected to the narrow slot of the receptacle. 2. A neutral side with two silver screws internally connected to the wide slot of the receptacle. 3. The green grounding screw internally connected to the receptacle's grounding slot. The electrician connects the black or red (hot) insulated wire of the incoming cable to the gold screw, which can be traced back to the main panel's circuit breaker. The white insulated (neutral) wire is connected to the silver screw, which can be traced back to the neutral bus on the main panel. The green screw is wired to the incoming ground wire, which eventually find its way to earth ground via the main panel. The black and white wires provide a current path to and from the load or appliance plugged into the receptacle. The grounding wire (called the equipment-grounding conductor) places the frame of the load at 0 volts, or ground potential, and provides a return path for the current to the panel if a ground fault occurs (see the drawing on the facing page).

If you have forgotten which cable is the incoming power, follow this trick of the trade: Always bring power into the box through one particular knockout—I prefer to bring power through the upper-left knockout. Doing so will enable you to tell with a quick look which cable is incoming and which cables are outgoing (see the drawing at left.)

Connecting the Wires to the Terminals

There is a correct way to insert a wire under a receptacle screw (or any screw for that matter) and an incorrect way. This may seem elementary, but it is important and most often done wrong by amateurs who end up leaving wires loose and arcing. Once stripped, the wire end must be twisted into a clockwise loop with needle-nose pliers. The loop cannot form a complete circle because it must be wound around a screw. Be sure all the looped wire is under the screw and not sticking out from it. As the

Which Cable is Which?

Incoming power

Outgoing power

Outgoing power

Outgoing power

If you always bring power into box via one particular knockout and run outgoing cables through others, you'll always be able to tell at a glance which cables are outgoing power and which one is incoming power (author uses upper left).

Receptacle Terminals

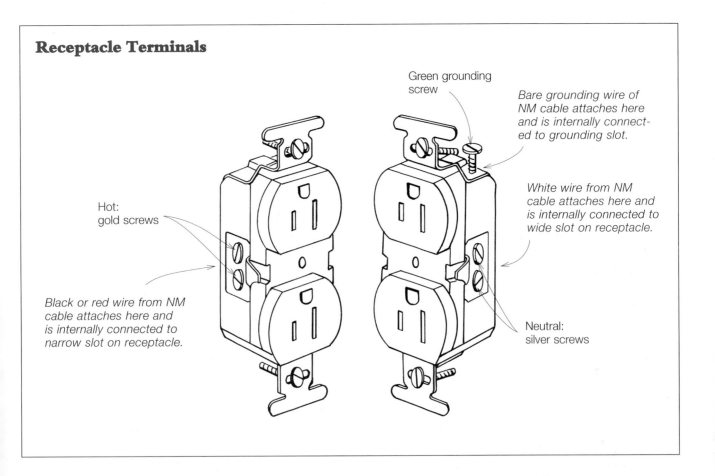

Green grounding screw

Bare grounding wire of NM cable attaches here and is internally connected to grounding slot.

Hot: gold screws

White wire from NM cable attaches here and is internally connected to wide slot on receptacle.

Black or red wire from NM cable attaches here and is internally connected to narrow slot on receptacle.

Neutral: silver screws

screw tightens in a clockwise direction, it will grab and pull the wire a little farther around the screw, tightening it between the screw and receptacle plate. Tighten securely, but not enough to strip the terminal's threads. If the threads do become stripped, throw the receptacle away. Once stripped, the most you can get is a loose connection that can cause the receptacle to overheat and eventually cause a fire.

There are many wrong ways to attach wires to the receptacle. Here's a list of what you *should not* do when wiring the receptacle:

1. Do not overlap the wire onto itself as it comes around the loop.

2. Do not leave a large part of the bare wire loop sticking out from the screw.

3. Do not wrap the wire around the screw in a counterclockwise direction.

4. Do not leave the wire in contact with only a small part of the screw (the wire must go at least three-quarters of the way around the screw).

5. Do not leave an unused screw sticking out from the receptacle. These can short out on metal boxes and unsuspecting fingers. Screw in all unused screws tight.

What to do with Extra Wires
Each screw on the receptacle can have only one wire under it. If you've got power coming in to the receptacle, and the receptacle feeds another two circuits, do not be tempted to twist two wires together and tighten them both under one screw. This is dangerous because the wires will eventually

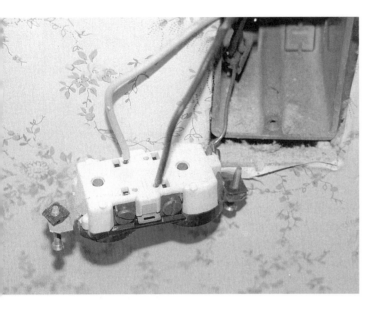

Although it may seem easier, don't attach the wires to the receptacle using the push-in terminals on the back. The connection could work loose, leading to overheating of the receptacle or to an intermittent circuit. Always use the screw terminals to attach wires to the receptacle.

work loose, resulting in an intermittent circuit, if the receptacle doesn't break apart in the first place.

You may also be tempted to put any extra wires in the push-in terminals—available on some receptacles—or you may find it easier to attach all wires in the push-ins. But don't do it (see the photo at left). I once had a service call in which the owners were having so many problems with the receptacles in the house, they thought the home to be haunted. When I got there and put my hand close to several of the receptacles that had nothing plugged into them, I could feel heat—enough to have melted the child-protector inserts. Other outlets were generating enough heat to deform the kitchen-appliance plugs. A few hours earlier, sparks flew out of an outlet in the nursery that had nothing plugged into it.

The problem was that the installer used the push-in terminals. Going by various trade names such as EZ wire, Quickwire and Speedwire, this method uses the spring tension of a metal "check" (similar to a clamp) inside the receptacle to hold the wire after it is pushed into a hole in the receptacle back. Such a system is custom-designed for a lazy electrician or do-it-yourselfer. It occasionally gives poor contact with the inserted wire (especially with small-diameter 14-gauge wire), leading to overheating of the receptacles or total loss of power to one or all receptacles in a string. In the interest of safety, never use the push-in terminals. Since January 1995, push-in terminals have been banned from 20-amp circuits. New receptacles will prevent the insertion of 12-gauge wire into the push-in terminals but will still be able to take 12-gauge wire under the screw terminals.

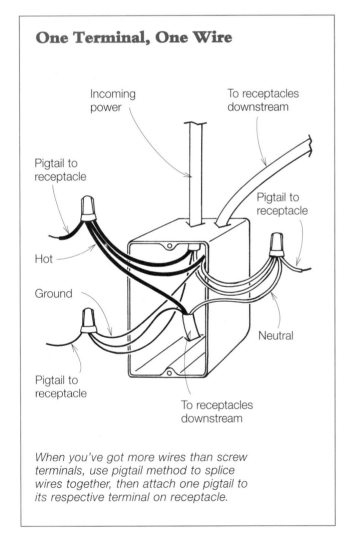

One Terminal, One Wire

Incoming power

To receptacles downstream

Pigtail to receptacle

Pigtail to receptacle

Hot

Ground

Neutral

Pigtail to receptacle

To receptacles downstream

When you've got more wires than screw terminals, use pigtail method to splice wires together, then attach one pigtail to its respective terminal on receptacle.

In my opinion, this is too little, too late: Push-in terminals should be banned altogether.

So what do you do with the extra wires when there are no extra screws and you don't want to use the push-ins? One method is to purchase a high-quality receptacle that accepts four hots and four neutrals using screw pressure on a metal bar to hold the wires (see the right photo on p. 146). Otherwise, the wires must be spliced together mechanically with an insulated cap or wire nut (for more on wire nuts and splicing, see pp. 152-154), with one short wire called a pigtail extending out from the splice to connect to the receptacle (see the drawing on the facing page).

Pushing Wires Back in the Box

Once wired, the receptacle should not be just shoved back into the box, making the wires bend and crumple in a haphazard manner. If only one receptacle and just a few wires are in the box, there shouldn't be a problem, regardless of what you do. However, if there are several wires and a couple of receptacles, there's a trick to getting all the wires back in the box without using a hydraulic jack: Have a plan. Be sure that all cable sheaths have been stripped off to ¼ in. to ½ in. of where they enter the box, as opposed to having several extra inches taking up box space. Next, cut off all excess lengths of wire; 6 in. from where the cable enters the box is plenty. Bring all the grounds to one side, neutrals to another, and the hots to still another; don't have wires threading through each other, making the box look like a bird's nest.

Once everything is hooked up, neatly push the wires into the box. First the ground wires: use your side cutters to push them up against the back of the box. Then push the neutrals to another part of the box. Then push the hot wires in the back, routing them along the edge of the box—not the center. The center of the box will need extra room for the receptacles. If the receptacles don't fit after all this,

you'll have to increase the size of the box or remove some cables.

Grounding a Receptacle in a Metal Box

To ground a receptacle, simply attach the bare ground wire to the green ground terminal on the receptacle. If you've got a metal box, however, you've got to do a bit more work. Metal receptacle boxes must be grounded, or they can become lethal conductors if a hot wire contacts them. Being grounded, the current from the hot wire will flow back to the main panel, causing the breaker to kick. This is done one of two ways. The most common is by attaching a bonding or grounding wire (a green screw connected to a 6-in. green insulated or bare wire) from the metal box to the grounding splice within the box (see the drawing on p. 162). You can make your own or buy them premade (I keep a green marking pencil to color any homemade grounding screws). Second, you can take a short bare wire from the grounding splice and attach it to the box via a metal push-on clip. Forget this method because the clip is hard to put on and is always popping off. The metal grounding clip you see on some high-quality receptacles is for automatic grounding of the receptacle, not the box.

If the wiring system has continuous metal conduit back to the service panel, the conduit itself can be the grounding path to ground the box and even the receptacle. This type of system lends itself to commercial and industrial use and is rare in residences. A danger in this type of grounding system is that if a piece of conduit gets loose or is removed (or replaced with nonmetallic conduit), the entire grounding system is lost.

Early wiring systems were without grounding wires, and many people died for lack of them. If your house has two-slotted receptacles, they're not grounded, nor will be any appliance that is plugged into them be grounded, even if you're using a plug-in adapter (also called a cheater plug). Such an

Grounding Receptacle in a Metal Box

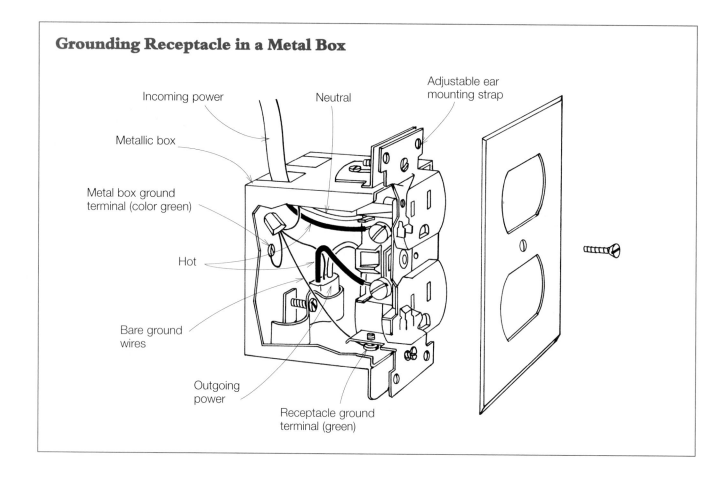

Incoming power

Neutral

Adjustable ear
mounting strap

Metallic box

Metal box ground
terminal (color green)

Hot

Bare ground
wires

Outgoing
power

Receptacle ground
terminal (green)

adapter only allows a three-prong plug to obtain power from a two-prong receptacle: It does not ground it. And if you decide to replace the ungrounded receptacle with a grounded receptacle so that you can plug in your new appliance, grounding is still not taking place. For grounding to take place, a wire, conduit or other grounding conductor must connect the receptacle grounding prong to the main panel's grounding system.

Do not assume the plug is grounded just because a grounded receptacle is in place. It's possible the grounding wire could be loose, broken or the feeder is spliced into an ungrounded wire. Do not assume proper ground until you've checked with a plug-in tester. Code allows replacing an ungrounded receptacle with a GFCI receptacle—but I don't recommend it in lieu of running a grounding wire.

Even though the GFCI does offer some protection against ground faults, the appliance still isn't grounded (for more on grounding, see Chapter 7).

Testing

Once the receptacle is wired and powered up, it should be tested with a plug-in tester to make sure the receptacle is wired correctly (see pp. 43-44). The basic tester can tell you if most of the wires are connected properly. A more sophisticated tester can also tell if there is a bootleg ground (neutral shorting to ground within a few feet) and can also electronically create a 15- and 20-amp load. The sophisticated tester has a retractable grounding pin that allows the unit to work on two-wire receptacles as well as three.

Receptacles in Series

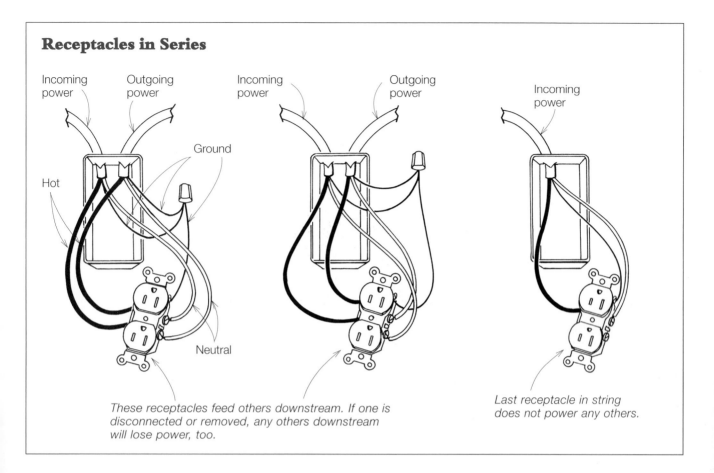

Incoming power

Outgoing power

Ground

Hot

Neutral

Incoming power

Outgoing power

Incoming power

These receptacles feed others downstream. If one is disconnected or removed, any others downstream will lose power, too.

Last receptacle in string does not power any others.

Receptacle Strings

There are two ways of wiring strings of receptacles together: a series string, like Christmas-tree lights, and a parallel string. The former is faster and cheaper but more problematic. The latter is the better system and the way I recommend.

Series string This string is the most common. It has the incoming current from the panel flowing through all the receptacles of the string. The disadvantage of this wiring setup is that if a wire comes off one receptacle, each receptacle that gets power through that receptacle will lose its power (see the drawing above).

Parallel string A much better system is to wire the receptacles in a parallel string, in which no current flows through a receptacle unless a load is plugged into it. Receptacles upstream can even be removed

without cutting off power to the others downstream. To do this the wiring must be pigtailed. With this system, the incoming black and white wires (the ones bringing in the power) are connected via wire nuts to the outgoing black and white wires that take power to the next receptacle. This takes the power directly in and out of the receptacle box without going through the receptacle itself. To get power to the receptacle, place 6-in. pigtail wires under the receptacle screws and take them to the two splices (see the drawing on p. 164).

Combination system What normally happens when electricians are wiring a house is a compromise of the two systems. As long as there isn't more than two black or two white wires to attached to the receptacle, the attachment is normally done under

Receptacles in Parallel

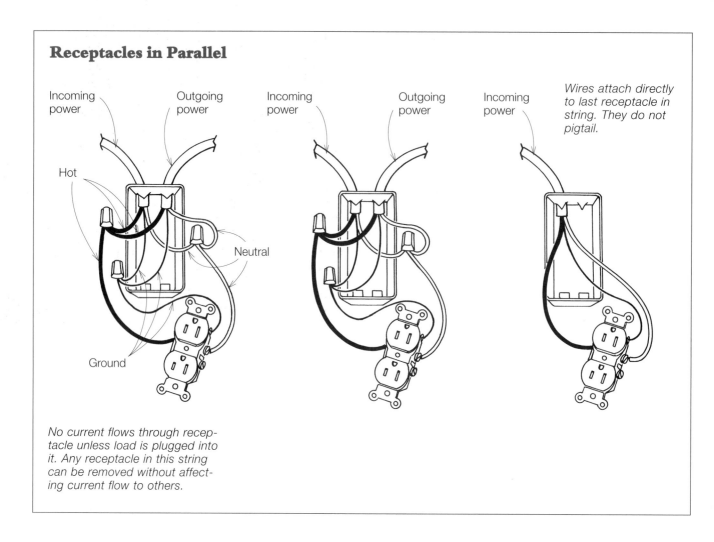

Incoming power

Outgoing power

Incoming power

Outgoing power

Incoming power

Wires attach directly to last receptacle in string. They do not pigtail.

Hot

Neutral

Ground

No current flows through receptacle unless load is plugged into it. Any receptacle in this string can be removed without affecting current flow to others.

the screws. If there is more than that, the wires are normally pigtailed. However, if you want the job done right, the way a good master electrician would do it, insist that *all* receptacles be pigtailed.

Troubleshooting

Solving problems is the most time-consuming part of the job. A good electrician will not only know what to do when things go right but also what to do when things go wrong. Here's a list of common problems.

Box has stripped-out receptacle-holding threads A common installation problem is to strip the box's female threads, so the receptacle's screws cannot pull the receptacle up to the box. This normally happens when you do not bend the wires back into the box;

rather, you depend on the screws to pull the receptacle into the box to bend the wires. However, this problem eventually happens to everyone—so what do you do about it? The most common screws lying around a building site are drywall screws, and these work nicely. They're easily obtainable and have a flat head so that the face plate will still fit tight against the receptacle.

Box's receptacle-holding threads have broken away What do you do if the entire female threaded portion of the box that holds the receptacle to it breaks off? Obviously, you can replace the box, which is not much of a problem if the finished walls aren't up yet. But if they are up, this can be very problematic. And besides, this only happens at the end of a long, hard day where everything else has gone wrong, and you have neither the time nor patience to change a box that is already surrounded by drywall. Rather than pull your hair out, try this: Take some adhesive glue, or other like material, and squirt some into the area where the screw will attach. After it hardens, predrill and use a drywall screw.

Wires won't fit into box Another common problem is having too many wires in the box and the receptacle doesn't seem to fit. Cable-fill violations aside, you can get more wires into a box by being neat and cutting away all excess sheath and wire.

Box was installed wrong If the receptacle box is installed improperly, you'll have a hard time installing the receptacle. If the receptacle box extends beyond the finished wall, so will the receptacle and the face plate, leaving an unsightly gap behind both. You will have to move the box in somehow or grind it down (this also grinds away the female screw attachment threads for the receptacle).

If the box is set too deep within the wall cavity, you will be in code violation (box must be flush with finished wall), and therefore the receptacle also will be set too deep in the wall cavity. For noncombustible surfaces such as tile, the box cannot be recessed farther than ¼ in. from the finished wall. If the drywall is cut perfectly around the box, the plastic ears (I call these Mickey Mouse ears) will affix to the drywall. But it's rare that you get a tight cut;

normally it's just the opposite. Spacers can be placed between the receptacle and the box to extend the receptacle out to the finished wall: I use small washers, nuts and Mickey Mouse ears with long attachment screws (always break off the little ears from any receptacle you're throwing away and save them; they make perfect washer-type spacers just for this purpose).

Some manufacturers make adjustable extensions for standard single-gang plastic boxes. They are hard to find, so you may have to check several stores. Beyond that, I cut the attachment nails with a reciprocating saw and pull the box forward, remounting it with drywall nails and glue. If the box is installed slighted canted to the left or right, receptacles have two horizontal slots, one above and one below, that allow you to plumb the receptacle.

Cut wires Sometimes the biggest headaches come from other people's work. On one job the drywall installers used a broken drill bit on their router to cut out each hole for the receptacle boxes. They would first attach the drywall, then plunge the bit through the drywall and into the box, cutting their way toward the box edge. Once they knew exactly where the box edge was, they inserted the router on the outside of the box and routed out the receptacle hole. They cut every wire in every box, including the telephone wires. It was a good thing I was being paid by the hour on that job. I now tell drywall crews and whoever is in charge, "Whoever cuts the wires pays for it."

Chapter 10

SWITCHES

SWITCH BOXES

SWITCH GRADES

TYPES OF SWITCHES

LIGHT DIMMERS

FAN SPEED CONTROLS

Switches control the power used throughout the house. For proper control, the right switch and box must be installed, and the switch must be wired correctly. Installing switches is not that difficult—wiring them is not, either. First you need to know what kind of load demands you will be placing on each switch. Then you must choose the box and switch that are best for that demand. It's also important to follow a system for wiring each switch.

In this chapter I will discuss types of switches—what grades are right for different situations—and how to wire them correctly. I'll show you how to make the job simpler by using some common sense and by following a plan. It all begins with choosing the right switch box.

SWITCH BOXES
The switch box is just as important as the switch. The wrong box design or size can not only cut and break the wires, but your setup could also fail the inspection due to overcrowding. Switch boxes are the same boxes used for receptacles. Therefore they have the same size, construction and volume limitations (see Chapter 9). But there are many companies that make and sell switch boxes as opposed to receptacle boxes: I advise you to stay away from them because these boxes tend to be small, which could easily place you in a cable-fill violation.

Because there are so many boxes to pick from, the initial challenge with switch boxes is knowing what type of box is needed for each switching situation. Normally, a single switch requires a single box—known in the trade as a single-gang box. Two switches normally use a double-gang box, three switches require a triple-gang box, and four require a four-gang box, or quad (see the photo at right). I don't recommend installing more than four switches at a single location, since nonmetallic boxes larger than four gang aren't available. In addition, switch face plates larger than four gang will be hard to find and may be expensive.

As an electrician, I often have to install a switch in a wall in which the studs are too close to allow the correct box size—I might need a triple-gang box but only a two-gang will fit. In this situation I use a double switch, which has two switches on the same yoke, or frame. If I cannot move the studs to fit the correct box in the wall, the double switch allows me to use a smaller box—a single instead of a double, a double instead of a triple. I also use double switches when I forget to wire in a switch, or when a design gets changed at the last minute. And because a double switch looks different than the other switches in the box, I can easily identify what each switch does.

Nonmetallic boxes with integral nails will be the most-used switch boxes in the residence. However, these boxes can be damaged easily and so must be kept within the wall cavity. Metal boxes will rarely be used within the wall cavity because of their higher cost and extra labor to install (you must ground the box). However, if the box is to be exposed, say on a concrete-block wall in an unfinished basement or garage, a metal box should be used.

Single-Gang Boxes

A single-gang box with one switch is the most common switching arrangement in a residence. Even though switches normally requires less volume in the box than receptacles do, you can easily put yourself in cable-fill violation if you purchase a box that's too small in volume. I never use a box smaller than 18 cu. in., and I prefer to use the bigger 20.3- and 22.5-cu. in. boxes that I use for receptacles—even though they are significantly more expensive than the 18-cu. in. boxes. As with receptacle boxes, nonmetallic switch boxes have their volumes

Switches use the same basic boxes as receptacles—but tend to be ganged more. Four types of switch boxes are shown here: from the top, single-gang box, double-gang box, triple-gang box and four-gang, or quad, box.

stamped inside; metal boxes do not. The chart below lists the most commonly used boxes, their volumes and the number of standard cables that can be placed within them, along with a switch.

A typical switching situation would have the following: a 12-gauge cable that brings in power to the switch; a 12-gauge cable that takes power to the load; a 12-gauge three-conductor cable for three-way switching; plus the grounding wires. Each 12-gauge cable requires 4.5 cu. in. of volume. The 12-gauge three-conductor cable requires 6.75 cu. in., and the grounding wires require 2.25 cu. in. Add up the

cubic inches required, and you'll know what box size you need. This situation would require a box with at least 18 cu. in. of volume.

Double-Gang Boxes

Switching via two switches in a double-gang box allows you to control power to two different loads from one location. Typical examples would be at the main entrance, where the outside lights are switched on at the same location as the interior lights, at a junction of a room and hall, and for controlling a ceiling fan and light from one box (see the drawing on the facing page).

Number of Cables Allowed in a Single-Gang Box

Single-gang box types and sizes	Number of 12-gauge cables with switch	Number of 14-gauge cables with switch
Metal handy box, 1½ - 1⅞ in. deep (11.5 - 13 cu. in.)	0	1
Metal handy box, 2⅛ in. deep (16.5 cu. in.)	1	2
Metal box with integral nails, 2½ in. deep (12.5 cu. in.)	0	1
Metal box with integral nails, 2²⁷⁄₃₂ in. deep (15.8 cu. in.)	2	2
Metal box with integral nails, 2⅞ in. deep (18.8 cu. in.)	2	3
Nonmetallic box, 16 cu. in.	2	2
Nonmetallic box, 18 cu. in.	2	3
Nonmetallic box, 20.3 cu. in.	3*	3
Nonmetallic box, 23 cu. in.	3*	4

*Cable-fill violation because 24.75 cu. in. of volume will be required.

Two Switches, Two Loads

This setup allows two loads to be controlled from one location. Each switch is fed from one cable, but each powers separate load.

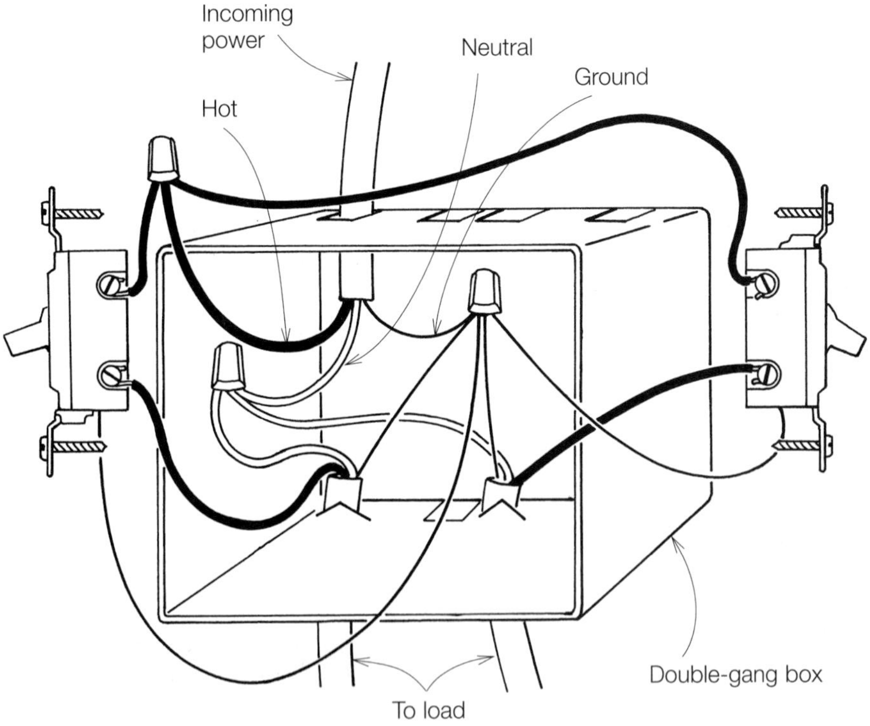

Incoming power

Neutral

Ground

Hot

Double-gang box

To load

Double-gang nonmetallic switch boxes have a variety of attachment mechanisms, depending on the manufacturer, and they are normally available in volumes of 25 to 34 cu. in. The smaller boxes should be avoided because they have barely enough volume to accept one incoming power cable and two load cables—the bare minimum required in a box with two switches. You will be in cable-fill violation as soon as you pull a cable off to feed a string of receptacles, or even if you have a three-way switch cable in the box. In addition, dimmers and fan-motor controllers occupy considerably more space in the box than simple switches, making it even more crowded in a small-volume box. The 34-cu. in. box is the minimum-volume box that should be used in double-gang switching.

Double-gang metal boxes can be used within stud walls, but it is rarely done because the boxes cost more and require extra labor to ground the box. However, a metal box is required on an unfinished basement or garage wall, and conduit will have to be run to protect the wiring at least up to 8 ft. overhead. In this situation, the box will need knockouts, as opposed to NM cable clamps. Many designs are available, but the particular models I recommend accept the switches directly into the box, just like nonmetallic boxes, and can use standard switch face plates. Use boxes that are 2⅛ in. deep if possible; 1½-in. deep boxes should only be used in a 1½-in. deep stud cavity (like a furred-out finished basement wall). Generally, the 1½-in. deep box can only have three 12-gauge cables.

Another metal box design has screw attachments only on its corners to accept special rounded metal face plates. The switch connects to the face plate, and the face plate is screwed onto the box. The face plate makes the box more expensive and makes switch installation rather awkward. The switch will have to be listed for single-screw attachment, and its ears will have to be removed so that it will fit in the box. Such a box design lends itself to splicing rather than to switch installation.

Triple-Gang Boxes

It is very common to see three switches at the main entrance of a house to have control over the outside light, the entry light and the interior light. Three switches (in a triple-gang box) is common whenever an overhead fan and fan light occupy two spaces in the box and a third is needed to switch the hallway light or another load. Triple-gang boxes typically are available in volumes of 45 and 54 cu. in. for nonmetallic boxes. I only recommend the larger box. You'd be surprised how quickly the box will fill with cable. Triple-gang nonmetallic boxes can be attached via integral nails or metal brackets.

Triple-gang switching via surface-mounted metal boxes is almost nonexistent in residential situations. Such boxes and covers are extremely hard to find and should not be designed in the electrical system unless the boxes are in hand. It's better simply to use double- and single-gang switching.

Four-Gang Boxes

A four-gang, or quad, box with four switches is the largest switching arrangement I recommend within a residence. If you must use a box larger than a four-gang, make sure the switches are clearly marked: It's hard enough to remember what four switches do, let alone additional ones. A four-gang arrangement could be used at the main entrance, controlling entry lights, room lights, spot lights and outside lights from one location. Or, if the hall opens between the dining and living rooms, you might want switching for a fan, a fan light, as well as lighting for the dining room, living room or the hall. The problem will be remembering which switch is which. If this situation arises, you might want to pull the two fan switches away from the others and install them in a double-gang box.

Nonmetallic boxes stop at four gang; above that, metal boxes will have to be used. Because boxes larger than four-gang are hard to find, try using several double-gang boxes instead. Typical volumes for nonmetallic four-gang boxes are 48, 58 and 73 cu. in. Always use the 73 cu. in. Four-gang boxes will need a strong attachment means—they normally will not come with integral nails but with another means of support. Some types have extendible arms for support, and others will need some sort of 2x bridging for additional support.

SWITCH GRADES

Switches, like receptacles, are available in many different grades—varying in safety and quality. The spectrum starts with the thin, low-cost, residential or general-use grade and ends with the high-quality but expensive spec grade.

The most common switch installed in a residence is residential grade, normally rated at 15 amps/120 volts AC (they normally are not available in 20 amp or higher). In theory, a residential grade can have full-rated current through the switch for loads such as fluorescent and incandescent lighting and inductive and resistance loads for heating. For motor loads it shouldn't be used for more than 80% of its rated current and not beyond the horsepower rating of the switch—normally ½ hp at 120 volts. In my opinion, a residential-grade switch shouldn't be used for anything that will pull current that's close to the switch's maximum rating.

Residential-grade switches are inexpensive, and you get what you pay for. It's not uncommon for this grade of switch to fall apart during installation or die an early death if a load is pulling close to the switch's rated current. Occasionally the side screws strip out in the body, even with just more than minimum pressure, and the cheap plastic body breaks apart as the screws tighten down. In addition, most cheap switches are designed to be wired via push-in terminals. Never, under any circumstance, use the push-in terminals: Always use the screws. I've lost track of the number of such connections that have pulled out of the switch and created an open

What to Look for in a Switch

Dual grounding options: grounding terminal or stainless-steel self-grounding clip

Virtually unbreakable nylon toggle

Insulating safety barrier

Heavy-duty, corrosion-resistant brass and nickel-plated yoke

One-piece, copper-alloy contact arm allows for excellent conductivity.

Break-off drywall ears

On side, large brass-screw terminals.

Large silver cadmium contacts (wherever electrical connection is made)

circuit. Another danger with a push-in terminal is that the wire can loosen over time. A loose connection increases resistance and can lead to arcing and burning. This type of cheap push-in terminal should not be confused with the high-quality switch connections that insert the wires from the back but use screws to tighten down on a bar that applies pressure on the conductors.

For those who want something other than minimum grade, spec-, commercial- and industrial-grade switches are available and are becoming more popular. These switches have thick bodies—about twice as thick as residential grade. They are also available in 20- and 30-amp ratings. A typical commercial-grade switch only costs $1 to $2 more than a cheap, residential-grade one. A little extra money increases the quality tremendously. For those extra couple dollars, you can get the following features: switch voltage increased from 120 to 277

volts; 20- and 30-amp designs that not only increase the allowable current but also allow horsepower ratings of the switch to increase from ½ hp to 2 hp; wired to side screws—no push-in terminals; nylon toggle to resist breakage; silver cadmium contacts for longer switch use; an internal neoprene rocker for longer switching wear.

However, because there are so many switches used within a residence, the cost of an extra dollar per switch can add up quickly. Few contractors are willing to absorb such a cost when you can't even see what you're getting. However, if building for a specific individual who may appreciate quality as opposed to bottom-line price, a wise contractor will offer the option (suggesting dimmers as well) to the owner-to-be for a fair increase of price. This way both contractor and homeowner will profit by the increase in quality.

Three Methods of Wiring a Single-Pole Switch

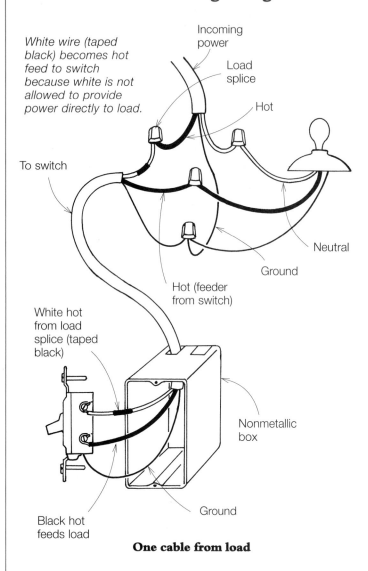

White wire (taped black) becomes hot feed to switch because white is not allowed to provide power directly to load.

Incoming power

Load splice

Hot

To switch

Hot

Neutral

Ground

Hot (feeder from switch)

White hot from load splice (taped black)

Nonmetallic box

Ground

Black hot feeds load

One cable from load

Incoming power cable (feeder) is brought directly to load, or light. Splice is made at load, not in switch box. From splice, one cable goes to switch. White wire to switch should be hot, not black wire, because black wire of same cable must be hot (switched) feed to load. (Adding black tape to hot white wires to indicate they're hot is not required but a good thing to do.)

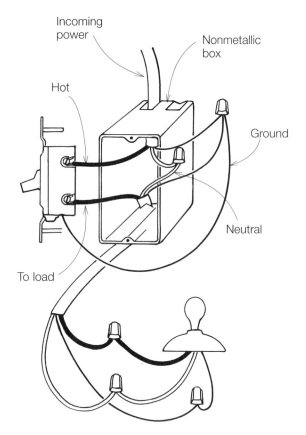

Incoming power

Nonmetallic box

Hot

Ground

Neutral

To load

Two cables: incoming power and load

Power is brought through black hot through switch, to load. White neutral wire from incoming cable is spliced directly to white from load. Ground wires from both cables are spliced together with ground from switch.

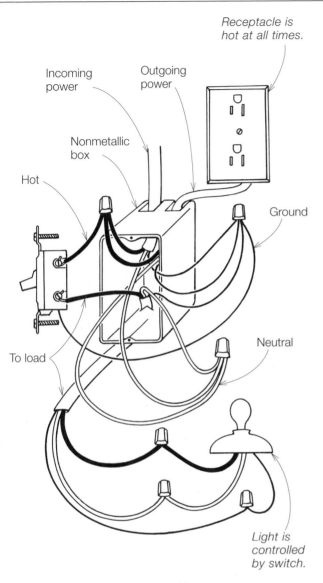

Incoming power

Outgoing power

Receptacle is hot at all times.

Nonmetallic box

Hot

Ground

To load

Neutral

Light is controlled by switch.

Three cables: incoming power, load, outgoing power

Receptacle downstream can be powered through switch box. Switch will control power to load, but receptacle will always have power.

A typical high-end design would be Bryant's Tech Spec switch, which is advertised as a super-safe design. It has a one-piece contact arm (no rivets to work loose), which allows better conductivity and less heat buildup; large silver cadmium contacts (where the actual connection is made) to reduce contact wear and to give a long life span; a large nylon handle (toggle) for ease of turning on and off and for longer wear; a neoprene rocker to keep the switch working after thousands of throws; and a heavy-duty yoke for mounting in the box. Bryant color-codes its switch bodies (not the handle that you see under the switch face plate) to make it obvious what current rating you're buying: blue for 15 amp, red for 20 amp and green for 30 amp. Handles normally come in ivory, brown, white, gray, red and black. Such switches normally come with a back wiring design, not to be confused with the cheap push-in terminals, that allow four wires to be attached (two under each screw clamp). This design often eliminates the need to pigtail, saving valuable labor time. So why doesn't everyone use such high-tech switches? Because each costs more than $10.

TYPES OF SWITCHES

There are many different types of switches, and your situation will dictate which type you choose (the most common switches in a residence are single pole, double pole, three-way and four way). The way switches have been wired has changed a lot since the turn of the century when push-button switches were used. Today, there are many more wires going to and from the switches, and the electrician must try to keep them straight. As with receptacle boxes, I normally run the incoming power cable into the switch box through the upper-left knockout, and loads out the bottom. This way you can tell if the power and load conductors are wired to the switch by just glancing at it. Even if you prefer to run the incoming cable through another knockout, be consistent.

Single-Pole Switches

A simple on/off snap switch is also referred to as a single-pole (SP) switch because inside the switch there is only one contact being opened or closed at a time. The switch allows electrical current to flow through it in the ON, or "make," position and opens the connection in the OFF, or "break," position. It's is the most common switch in a residence and is used exclusively to turn lights on and off from one location. The most common mistake in installing a simple switch is installing it upside down, making up OFF and down ON.

An SP switch has two terminals or screws: one terminal for the incoming hot wire and one for the wire called the "return," which takes the power to the load when the switch is ON. As far as making the circuit work, it makes no difference which wire connects to which terminal. But I prefer to follow a system: I always take the incoming hot wire to the top terminal, and take the wire to the load off the bottom terminal. This way I can tell at a glance what wire has been run to the switch. It also is an extremely good troubleshooting aid.

How you wire an SP switch depends on which box the power cable enters: the switch box or load box. Basically, there are three situations: power comes to the load box and then a wire takes it down to the switch; power comes from the panel directly to the switch box, then via a cable to the load; power comes from the panel to the switch box, to the load, and then an outgoing cable feeds a receptacle downstream. The drawings on pp. 172 and 173 illustrate how an SP switch is wired in each of these situations.

Wiring multigang SP switches is identical to single-pole wiring: You just have more wires and switches to deal with. Simply pigtail off the main black hot wire to each switch and connect all neutrals together. (Do not use the push-in terminals behind the switch as jumpers from switch to switch.)

Double-Pole Switches

A double-pole (DP) switch can turn two independent circuits on/off at one time because two contacts, or poles, are being opened and closed at one time. A DP switch is wired as two independent SP switches—one on the left and one on the right. In this situation the white is hot, not neutral, and must be taped black to

Wiring a Double-Pole Switch

240-volt incoming power

Ground

White is hot and must be taped black to indicate it's hot.

Hot

Switched power out

Double-pole switch controls 240-volt appliance and uses both legs of circuit. Its most common use is as cutoff switch for appliances that cannot be unplugged for maintenance, such as a submersible pump.

indicate that it's hot. A DP switch is normally used to control a 240-volt appliance, such as a submersible pump or other appliance that does not have a plug that can be disconnected for maintenance. The appliance therefore must have a cutoff switch adjacent to it or its controls to protect the person performing the maintenance. I wire a DP switch the same way I wire an SP switch: hot to the top, and load to the bottom (see the drawing above).

Three-Way Switches

A three-way switch controls a load from two locations, such as at the top and bottom of a staircase, or at either end of a hallway. A three-way

Wiring a Three-Way Switch

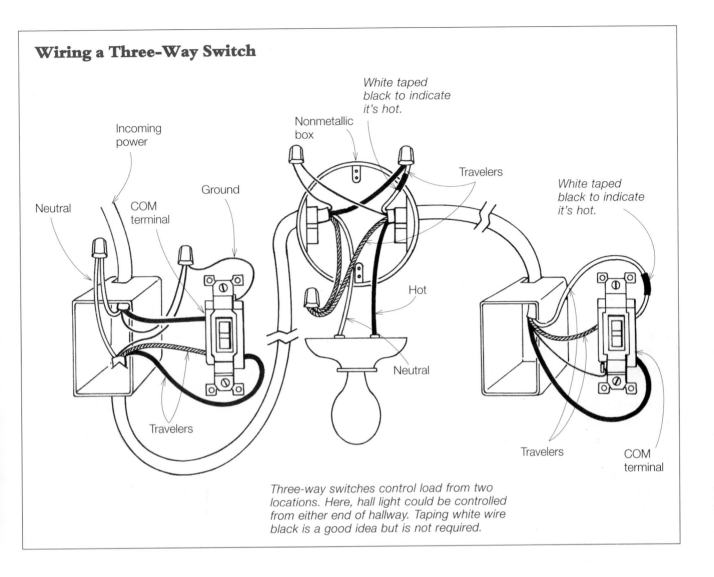

Neutral

Incoming power

COM terminal

Ground

Nonmetallic box

White taped black to indicate it's hot.

Travelers

White taped black to indicate it's hot.

Hot

Neutral

Travelers

Travelers

COM terminal

Three-way switches control load from two locations. Here, hall light could be controlled from either end of hallway. Taping white wire black is a good idea but is not required.

switch has three terminals. A COM (common) terminal noted by a dark-colored screw, and two terminals for the hot traveler wires that have no polarity (the switch has no ON or OFF). The COM terminal is also called the tongue to explain how the switch works: It laps up to connect to one traveler terminal and down to connect to the other.

Here's the simplest way to wire a three-way switching circuit (see the drawing above). Run the hot wire to the COM terminal of the first three-way switch. Run the neutral (of the cable with the hot wire) to the white load wire. Run the black load wire to the COM terminal on the second three-way switch. From the two leftover terminals of the first switch, run the two traveler wires (no polarity involved) straight to the

two leftover terminals of the second switch. Always use three-way switch cable, not standard two-conductor cable, and run the cable from the first switch to the load and then to the second switch.

Four-Way Switches

For additional points of load control between three-way switches, four-way switches are available. There is no limit to the number of four-way switches that can be installed, as long as they are installed between the three-way switches—a three-way switch must always be on each end of a multiple-point switching circuit. Four-way switching would be used if a room had three or more exits or entrances with switched lighting at each one.

Four-Way Switching

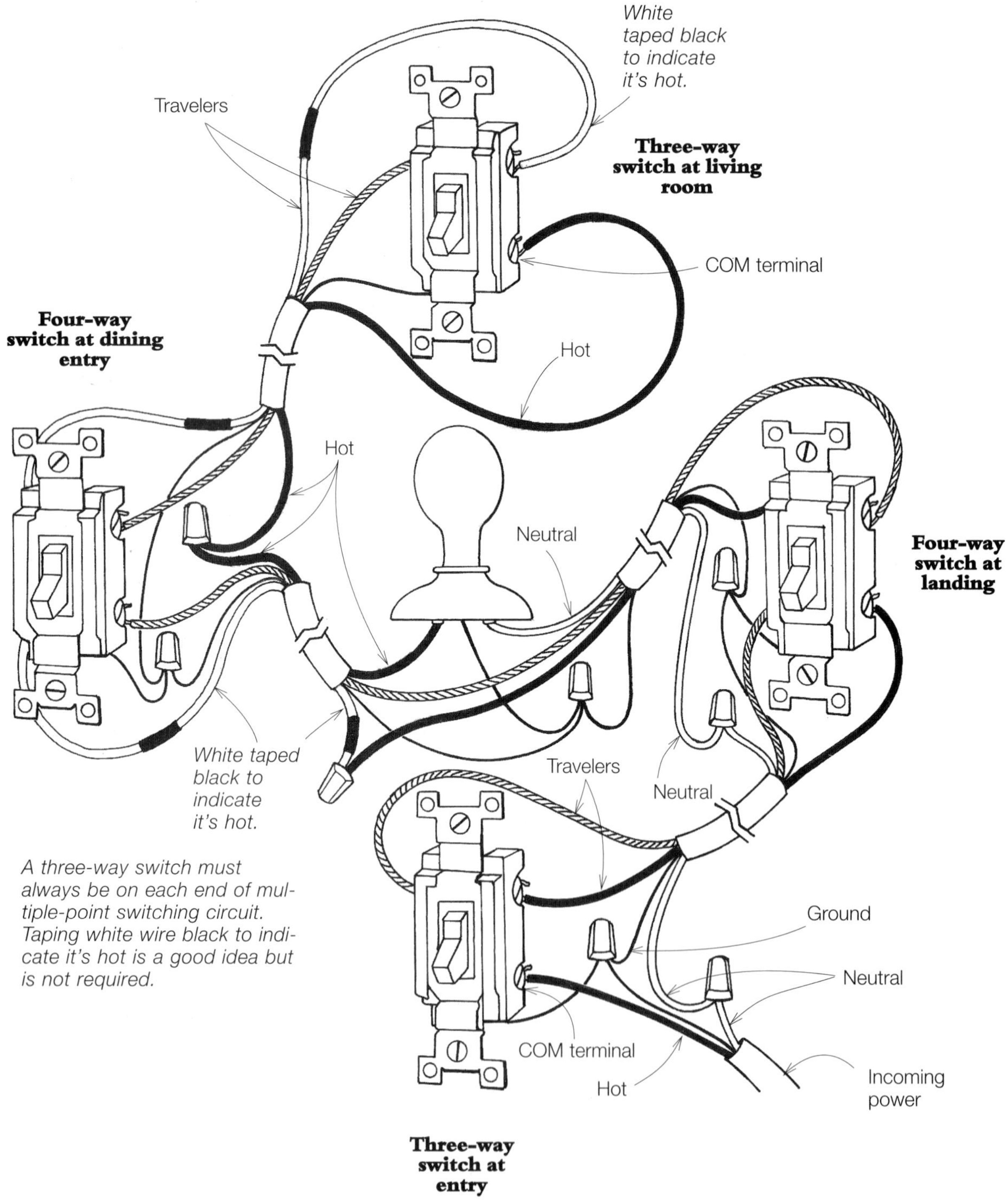

White
taped black
to indicate
it's hot.

Travelers

**Three-way
switch at living
room**

COM terminal

**Four-way
switch at dining
entry**

Hot

Hot

Neutral

**Four-way
switch at
landing**

White taped
black to
indicate
it's hot.

*A three-way switch must
always be on each end of multiple-point switching circuit.
Taping white wire black to indicate it's hot is a good idea but
is not required.*

Travelers

Neutral

Ground

Neutral

COM terminal

Incoming
power

Hot

**Three-way
switch at
entry**

The box the switch comes in may have a schematic on how to wire the switch. Four-way switches are easily wired incorrectly, so use a continuity tester to verify the switching arrangement (see the drawing on the facing page). I connect the incoming travelers to the bottom switch terminals and the outgoing travelers to the top terminals (verify the terminal connections). Attaching each set of travelers on one side of the switch will throw a short on the circuit.

Specialty Switches

Switches are available for just about any switching condition that exists. Lighted toggle switches that glow when off are for switches that are in dark locations or even in a child's room so the child will always know where the light switch is. Pilot-light switches that glow when the load is turned on are very useful when a switch is at a distant location from the load. The pilot light glowing is a reminder that the light or load is still turned on. Keyed switches are available for situations that require some type of security. Timer switches are available for temporary loads, such as a fan in the bathroom. Programmable timer switches are available to turn on lights, a radio or a television at preset times so that people will think someone is home, even when the house is empty.

Rocker switches, considered as a fashion design, are available from most manufacturers, with each having its own trade name. Different grades, amperage ratings, colors and specialty designs are available just like standard switches. However, rocker switches are considerably more expensive than standard switches. Special plastic, nylon or metal wall plates will also have to be purchased.

LIGHT DIMMERS

Dimmer switches were developed to allow more control over lighting other than simple ON and OFF. Originally, dimming was achieved via rheostats that were big and bulky and also wasted energy. Today's 600-watt electronic lamp dimmers were developed in the early 1960s. You can now get dimmers for both incandescent and fluorescent lighting.

Dimmer switches are available in single-pole or three-way models. High-tech electronic multilocation switching is now on the market and does allow for dimming at a number of locations from one central dimmer switch. The central unit is the master unit, and all other controllers are slaves to this unit. If any switch on the system goes out, it must be replaced with the same type of unit for the system to work properly. You cannot replace the slave or master with a standard mechanical control. In addition, such units require special wiring and need to be designed in the electrical system ahead of time. Some dimmer models have wireless remote controls, and some designer series can have up to 14 different color face plates or solid metal finishes.

How Dimmers Work

Today's dimmers use a solid-state switch called a triac that switches current on and off the load 120 times per second—twice the line frequency (see the drawing on p. 178). Dimming is controlled by limiting the amount of time that current flows to the load. The longer the current is off, the more the light is dimmed, with only about 2% of the energy wasted as heat (see the drawing on p. 179). The switching is done so rapidly that it cannot be seen by the human eye—only the average brilliance will be noticed. **Warning:** Dimmers should never be used to control a receptacle or any device that has a motor. If the load plugged into the receptacle doesn't get its full power, it could ruin the appliance and even start a fire. A fan or any motorized device needs its own specially designed control.

Today's dimmers save electricity by limiting the power to the load when full wattage isn't needed. In addition, they extend the life of incandescent bulbs by taking the initial surge off the line that burns through a cold filament and by reducing the heat within the bulb. This is especially important if the bulbs are hard to reach. If the lighting load is significant, the dimmer can pay for itself within a year. Perhaps the most significant use of a dimmer, other than energy conservation, is to use illumination to enhance architecture and landscaping, to accent paintings and other objects of art, and to create the proper lighting atmosphere

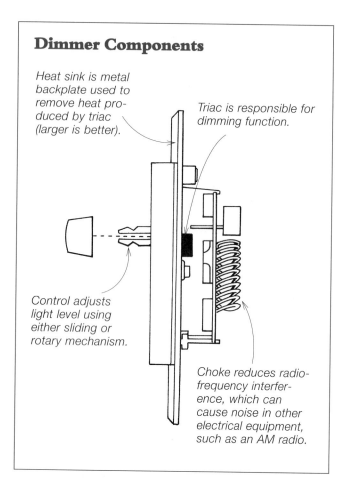

Dimmer Components

Heat sink is metal backplate used to remove heat produced by triac (larger is better).

Triac is responsible for dimming function.

Control adjusts light level using either sliding or rotary mechanism.

Choke reduces radio-frequency interference, which can cause noise in other electrical equipment, such as an AM radio.

going to be used (see the drawing on p. 180). If there is to be only one switch, the two wires that would have gone to the switch simply connect to the dimmer. Always be sure power is off when installing a dimmer. If a three-way system is going to be dimmed, a decision will have to be made on which switch side the dimmer is going to be installed. If a four-way or larger system is going to be installed, the logic is the same—only one dimmer in the system. Dimmers with two black wires have no polarity, so it doesn't make any difference which wire gets hot or neutral. If the dimmer has a black and red wire, the black normally goes to the incoming power source, and the red to the load.

Fluorescent Dimmers

You cannot dim standard fluorescent lights because their ballasts are not designed for dimming. Instead, special electronic ballasts or magnetic dimming ballasts are required. Magnetic dimming ballasts work with T-12 fluorescent bulbs. This type of ballast uses large magnetic components that may emit a buzzing noise, and the lamps may visibly flicker at low light levels.

Electronic dimming ballasts are required for rapid-start fluorescent lights. Operating at a higher frequency, these ballasts eliminate both noise and flickering and offer an energy savings over the magnetic ballasts. Electronic ballasts must be matched to the voltage, wattage and length of the lamp (bulbs that contain diodes may not dim properly). If the unit is purchased new, everything is matched at the factory. Fluorescent dimmers have several application notes that you should research before installing them. For example, for best performance, you should check with the manufacturer before you mix lamp lengths or colors on the same dimmer circuit. Do not mix ballasts or lamps from different companies. Also, new lamps may have to be burned in at full intensity for 100 hours before dimming to achieve optimal dimming performance, and dimming control under certain situations may only be available down to 20%.

within a room. I personally like dimmers for the bedroom and bathroom lights to reduce their brilliance in the morning or in the middle of the night until my eyes readjust to normal light levels.

Wiring Dimmers

Single-pole dimmers can be installed by simply replacing a single-pole switch. However, three-way dimmers cannot simply replace every three-way switch. Only one of the two three-way switches can be replaced with a dimmer; the other must remain a standard three-way switch.

If standard dimmers are going to be used, the rough-in wiring can be done the same as if switches were

When designing fluorescent dimming into your plan, the two most crucial issues to address are the dimming range required (full range or partial) and the noise sensitivity of the area.

Common Problems with Dimmers

No product comes problem-free, and dimmers are no different. A good electrician will know the disadvantages as well as the advantages of dimmers so that he or she can install the products properly as well as make recommendations to the owner/builder.

Noise Dimmers are noisy—sometimes it's noticeable, sometimes not. They put radio-frequency interference (RFI) on the power line on both sides of the dimmer circuitry—the load and the entire branch circuit—by both conduction and radiation. Any load on the dimmer branch circuit will receive this noise. For example, you may hear it through AM radios, stereo equipment, broadcasting equipment, intercom systems, public-address systems and cordless telephones.

There are several ways to eliminate or reduce this noise. The first option is to buy a dimmer with a built-in noise filter. Next, though not required by code, it is always best to have the lighting branch circuits independent of receptacle branch circuits so that the dimmer does not emit noise through any sensitive equipment plugged into the receptacles. Bring power in to the switch from other light-only circuits or a separate branch circuit. It's also important to make sure all ground connections are tight. A poorly grounded or ungrounded light on a dimmer circuit will generate noise like a radio antenna. Running the dimmer wiring in its own well-grounded conduit may help to a limited degree.

Yet another option is to install a lamp debuzzing coil (LDC). An LDC is the most effective way to reduce RFI. You will need one for each dimmer, selected according to the load. Though an LDC stops noise through other equipment by suppressing RFI, it has its own audible buzz, so locate it where the buzz will not be objectionable. For maximum RFI suppression, keep the wiring between the LDC and the dimmer as short as possible. LDCs fit in a 4-in. by 4-in. junction

How a Dimmer Works

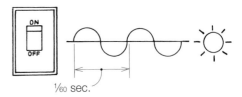

Full current to load

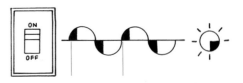

¾ current to load

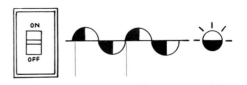

½ current to load

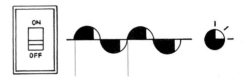

¼ current to load

Dimmer uses internal switching device (triac) to control light output. As external switch moves more toward ON direction, triac allows more current to flow to light.

Wiring a Simple Dimmer

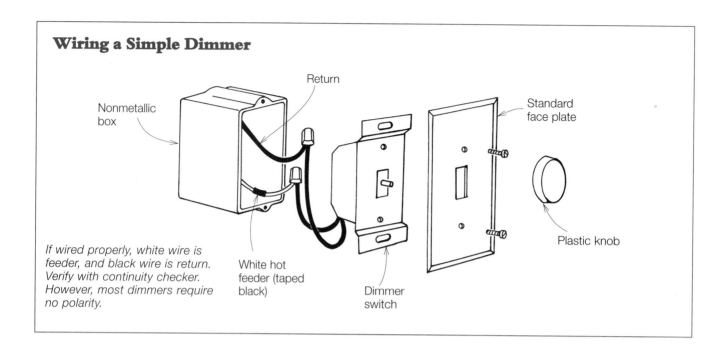

Nonmetallic box

Return

Standard face plate

Plastic knob

If wired properly, white wire is feeder, and black wire is return. Verify with continuity checker. However, most dimmers require no polarity.

White hot feeder (taped black)

Dimmer switch

box, are thermally protected and are UL-listed. Some dimmers emit an audible noise when dimming at low levels. Look for dimmers that are advertised as being quiet.

Lamp buzz As the current to the lamp is being adjusted, the lamp's filament will buzz (it's more noticeable with bulbs above 100 watts). All lamps buzz to some extent, but the problem can be minimized by installing high-quality light bulbs. There is a difference in filaments—for instance, rough service bulbs have a heavier filament that reduces or eliminates the vibration. LDC coils can reduce lamp buzz, and so will physically smaller bulbs.

Heat and overloading This problem is normally created by designers or homeowners who create lighting systems that pull too much current through the dimmer, destroying it with excessive heat. The solution is simple: Don't exceed the dimmer's wattage rating. This problem is extremely common when several light banks are ganged together, or when track lighting is involved. If the bank's bulbs get changed out for a higher-wattage bulb, or extra lights get added to the track lighting, the dimmer

can easily be overloaded. When this occurs, the triac inside the dimmer will overheat and become an open circuit. The dimmer will either simply turn on and off with no control over the dimming, or it simply will not work at all.

A standard low-wattage dimmer uses its aluminum mounting plate as a heat sink. Because heat is the bane of dimmers, the thicker this heat sink is, the better. Higher-wattage units, such as 500, 1,000 and 2,000, will dissipate heat through heat sinks mounted at the wall plate. If dimmers are ganged together in the same box, some of the fins on the mounting plate may have to be broken off so that they can overlap, reducing the size of the heat sink (see the drawing on the facing page). If this is the case, the wattage of each unit should be reduced by 20% to 50%—for example, a 1,000-watt dimmer would be reduced to a wattage of 800, or eight 100-watt lights. A UL requirement for solid-state wall-mounted dimmers is that the dimmer temperature measured on the heat sink is not allowed to exceed ambient room temperature by more than 150°.

Dimmer Heat Sinks

Aluminum mounting plate on standard dimmer serves as heat sink. When placing dimmers side by side in one box, part of plate may have to be removed, reducing size of heat sink. This will reduce wattage of lamp that dimmer can dissipate by 20% to 50%.

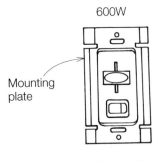

Mounting plate

No side sections removed, so dimmer operates at 100%.

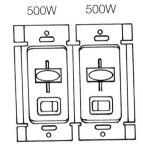

One side section removed from each dimmer.

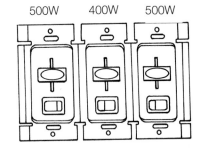

Middle dimmer has two side sections removed. Side units have one section removed.

FAN SPEED CONTROLS

There are two types of fan speed controls: fully variable, also called full range, and step control. Both types of controls may be used on a paddle fan, regardless of the number of speeds the fan has—simply set the fan's speed on its highest setting as it's installed. The wall-mounted speed controller can then vary the fan through all its speeds. Be sure to read the directions that come with the fan. A manufacturer may indicate a specific control to be used with its fans.

Noise is a problem with fans as well as with dimmers. I remember the first variable-speed control I installed and how loud the humming noise was when the fan was on. I've been trying to find quiet controls ever since. Step controls normally do not hum, but they can be used with only one fan, and the maximum operating current is 1.5 amps. Fully variable controls are noisier but have the advantage of being able to control the full-range speed of a single fan or multiple fans wired in parallel. The maximum current depends on the model, but 2- to 12-amp models are pretty common. As fan speeds reduce, the fan itself can emit an annoying hum. Be sure the controller addresses that issue as well.

Chapter 11

WIRING FIXTURES

Light fixtures help make a house a home, and each fixture has advantages and disadvantages that need to be known before it is installed. Again, planning is important. Even though fixtures are installed after the finished walls are up, you should know what kind of fixtures you'll be installing beforehand so that you use the correct rough-in ceiling box.

CHOOSE THE RIGHT BOX

There are many different types of ceiling boxes to pick from (see the drawing on the facing page). In general, ceiling boxes are round—not rectangular like receptacle boxes. Old-style boxes are 3½ in. in diameter, while the newer-style boxes are 4 in. in diameter. A ceiling box is screwed or nailed directly onto a joist (or between them). A fixture that weighs less than 50 lb. can be supported by a ceiling box alone. But a fixture that weighs more than 50 lb. must be supported independently of the box (see the drawing on p. 190).

There are several basic types of ceiling boxes. Standard cut-in boxes are made for renovation work, but these types of boxes are limited to fixtures that are very light in weight. Bar-hanger boxes are commonly used in new construction and renovation work. This type of box has a 16- or 24-in. bar that attaches between ceiling joists, and you can slide the box to any location along the bar. However, it tends to sag with even moderate weight. The more common fixture outlet boxes have captive nails or brackets for easy installation. These can attach directly to the ceiling joists or to a wood crosspiece between the joists for a more exact location. Typical

Ceiling Boxes

Nail-in boxes with brackets

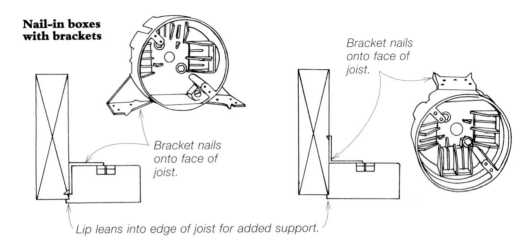

Bracket nails onto face of joist.

Bracket nails onto face of joist.

Lip leans into edge of joist for added support.

Bar-hanger box

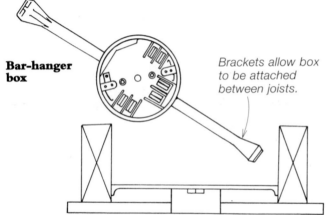

Brackets allow box to be attached between joists.

Box should not be used with heavy fixture because it might sag.

Integral nail-in box

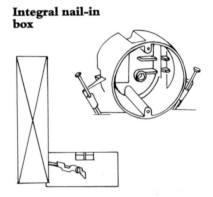

Nail-in boxes are inexpensive but break easily during installation.

Cut-in boxes

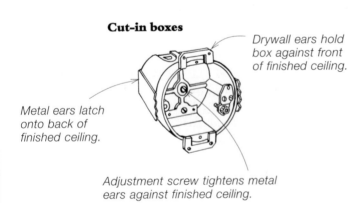

Metal ears latch onto back of finished ceiling.

Drywall ears hold box against front of finished ceiling.

Adjustment screw tightens metal ears against finished ceiling.

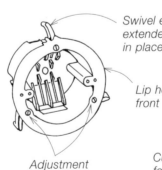

Swivel ears, once extended, lock box in place.

Lip holds box against front of finished ceiling.

Adjustment screws

Cut-in boxes are used for renovation work.

Pancake Boxes

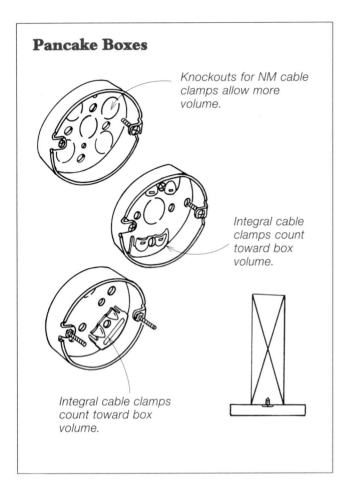

Knockouts for NM cable clamps allow more volume.

Integral cable clamps count toward box volume.

Integral cable clamps count toward box volume.

---

ceiling boxes, regardless of design, have volumes ranging from around 13 cu. in. to 23 cu. in.

A pancake box is one of the most common boxes. Pancake boxes are normally 3¼ in. to 4 in. in diameter, with volumes ranging from 4 cu. in. to 6 cu. in., and are normally only ½ in. thick (see the drawing above). The light fixture attaches directly to it—there is no cover plate on the box. When using such a box, be sure that you do not create a cable-fill violation, which is easy to do because of its shallow thickness and low volume. Pancake boxes are commonly used to support light fixtures with domed canopies, usually seen on the outside of a house adjacent to entries and exits. If volume is

a problem, use a box with knockouts for the cable, not a box with cable clamps built in, to gain a bit of volume. A pancake box is also used to flush-mount light fixtures on a wall, or on exposed beams, like in a timber frame.

INCANDESCENT LIGHTS

I do not care for incandescent light fixtures that totally encase the bulb. In my opinion, there must be a way for the heat generated by the bulb to escape, or the heat will be transferred into the light fixture and wiring. Most standard incandescent overhead light units will be 60 watts maximum per bulb. If you need more illumination, look for a light fixture that is UL-rated for a higher wattage. In addition, look for a light that is easily disassembled for bulb replacement. Standard bulb bases are better than the smaller, fancy, specialized bases—they seem to be sturdier.

Rough-in is simply mounting a standard ceiling box and inserting the cable to power the light. The light will mount via the two screw terminals on the ceiling box, by a special bracket that comes with the fixture or through the box and into a joist or 2x6 bridging. Power enters via a black or sometimes red wire and flows to the brass terminal on the bulb socket. Power then flows to a small tab in the center of the bulb socket, through the filament inside the bulb and finally out via the screw shell terminal and into the white neutral wire (see the drawing on the facing page). A ground wire connects all noncurrent-carrying conductors to ground.

FLUORESCENT LIGHTS

Fluorescent lights give a lot of illumination for the money and are very popular because of it. Standard fluorescents normally provide sufficient illumination, but if you require more, you can buy high-output and very high-output fluorescents. Typical problems with fluorescent fixtures in general is that they sometimes flicker and give off an annoying buzzing

Wiring an Incandescent Light Fixture

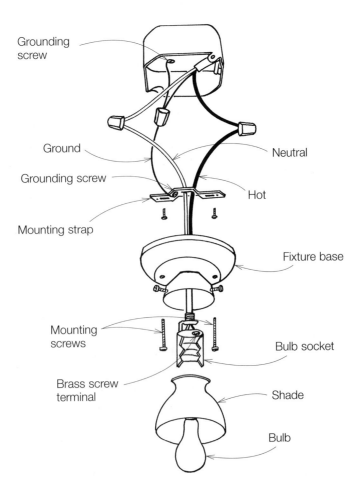

Grounding screw

Ground

Grounding screw

Mounting strap

Neutral

Hot

Fixture base

Mounting screws

Bulb socket

Brass screw terminal

Shade

Bulb

noise. The only solution is to change the ballast, which sometimes doesn't work, or install a very-expensive electronic ballast. Standard ballasts sometimes have a short life span and may short cycle when they get hot, cutting out the lamp until the ballast cools down. Standard ballasts are not meant for cold weather, either. If you plan to install fluorescent fixtures in an unheated shop, garage or basement, you may have to buy cold-weather ballasts, which are expensive. Corrosion is a problem in damp areas such as a garage or

basement. Both the male pins on the bulb and the socket will corrode over time and give intermittent problems.

Fluorescent lights come in either bare-bulb or enclosed fixtures, and which type you choose will be based on aesthetics: The enclosed fixtures are more attractive than the bare-bulb fixtures, but the plastic enclosure will turn yellow over time and may have to be replaced. The most common fixture lengths are 4 ft. and 2 ft., and these fixtures take

Wiring a Fluorescent Fixture

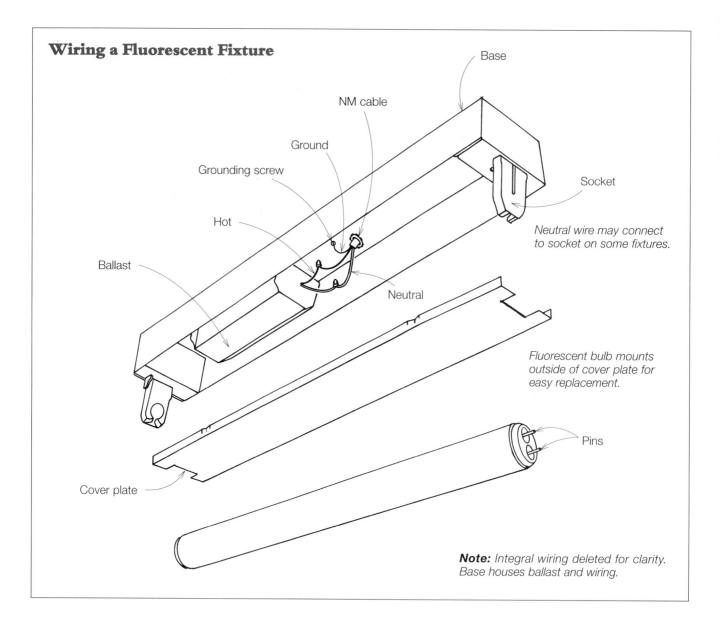

Base

NM cable

Ground

Grounding screw

Socket

Hot

Neutral wire may connect to socket on some fixtures.

Ballast

Neutral

Fluorescent bulb mounts outside of cover plate for easy replacement.

Pins

Cover plate

Note: *Integral wiring deleted for clarity. Base houses ballast and wiring.*

bulbs with two pins on each end (see the drawing above). For larger rooms, you might consider installing 8-ft. fixtures. The bulb for an 8-ft. fixture has a single pin on each end, which makes it much easier to change compared to a two-pin bulb. But they can be unwieldy.

Some of the cheaper designs on the market utilize a single-bulb, half-width base, which may only be 1 in. wide. The small base makes it very hard to cover the hole in the finished wall for the cable. Others are so narrow that wiring within them is impossible, and

they must be attached somehow to an electrical box. Don't buy these cheap fixtures unless you have a specific use for the small design.

Installation of fluorescent fixtures is easy. For a hard-wired unit, simply take the NM cable through a knockout in the box (use an NM connector) and screw the fixture into the ceiling joists. Wiring the fixture is simple: black to black, white to white and ground to frame. Some fixtures are equipped with a plug and cord. These units will require you to install a standard 120-volt receptacle (see Chapter 9).

Never attach a fluorescent fixture directly to the grid frame of a drop ceiling. The fixture will crash to the floor if the ceiling frame bends even a little (I've seen it happen), and anyone standing under the fixture could be severely injured. Fluorescent fixtures are heavy and should be independently and securely mounted to ceiling joists or 2x6 bridging.

Don't be conned into installing the standard fluorescent bulbs, such as "warm white" and "cool white," that are commonly available. Color rendition is important. To me, warm white is a depressing yellow, while cool white is a little better but not much. The best fluorescent light bulb is called "daylight." It's a little more expensive than a standard bulb, but it's worth it. Daylight bulbs are available for most fluorescent fixtures.

Under-Cabinet Kitchen Lighting

Kitchen lighting for the countertop is done quite easily via fluorescent lights mounted under the kitchen cabinet. Such illumination removes the shadows that are invariably present when using ceiling light. Try to get a fixture that is narrow—about 1½ in.—so that it can be hidden. Such a lamp normally has a built-in switch. Some can be hard-wired, and others come with a plug and cord. A plug-and-cord fixture will require a 120-volt receptacle nearby. For a hard-wired unit, pull the rough-in wiring out just under the kitchen cabinets—or even better, just above the cabinet bottoms. If you're designing a new kitchen that will have under-cabinet lights, try to install cabinets that have at least a 1½-in. lip on the outside to hide the light. As the cabinets are installed, cut the drywall for the cable so that the cable just fits under the cabinet's bottom.

CEILING FANS

Ceiling fans have been very popular since Hunter started making them in the late 1800s. But before you think about installing one, you'll need to consider a few things. First, you must be sure that you have enough overhead room. Don't put a ceiling fan in a bathroom, where people will raise their arms to undress or dry off. You'll need at least 7 ft. of clearance from floor to blades; less than that, and you'll be giving people crewcuts as they walk by (see the drawing below). For a low-ceiling room, you can buy a special kit that allows the fan to be installed flush to the ceiling. And you'll also need

Fan Clearances

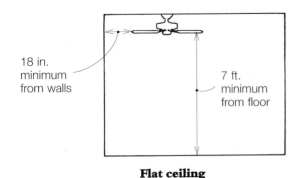

18 in. minimum from walls

7 ft. minimum from floor

Flat ceiling

Keep fans minimum 7 ft. from floor. For ceilings higher than 8 ft., position blades 8 ft. to 9 ft. above floor.

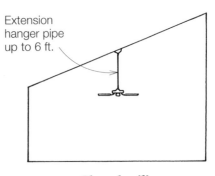

Extension hanger pipe up to 6 ft.

Sloped ceiling

For sloped or cathedral ceilings, use mounting kit with extension hanger pipe to achieve necessary clearances.

Sizing a Fan

Use 52-in. fan for rooms up to 400 sq. ft.

Use 44-in. fan for rooms up to 225 sq. ft.

Use 42-in. fan for rooms up to 144 sq. ft.

Use 32-in. fan for rooms up to 64 sq. ft.

These are guidelines for rooms with 8-ft. ceilings. Larger rooms may require more than one fan.

at least 18 in. from any wall to the ends of the blades for efficient air movement. Another important consideration is the ceiling pitch. The fan will have to hang far down enough that its blades don't hit the ceiling. Also, I don't recommend buying a ceiling fan with light fixtures, unless the manufacturer guarantees that the fan will not wobble—and most fans wobble. And if the fan wobbles, so will the lights. If you do install a ceiling fan with lights, make sure the lights don't hang so far down that they can be hit by a person's head or raised arms, which can break the lights easily.

Fans range from about 32 in. to around 52 in. in diameter. To get the most air movement and comfort, buy the appropriate-size fan for the room in which it will be installed (see the drawings at left).

Buying Guidelines

Try to buy a fan that's designed intelligently. Make sure the fan will allow you to hang it at an angle while you wire it, and then, once it's wired, swing it up into place. Nothing is worse than trying to wire a fan that has no swing-up bracket. In the old days I would have to hang the fan from a homemade rig while I wired it and then remove the rig to install the fan once I had the splices made. Here are some other things to look out for:

• Low-cost fans have a cheap motor. Therefore, to compensate for the low-power motor, the pitch of the blades has to be lessened (made more parallel to the floor) so that it doesn't overwork the motor. As a result, not much air gets moved.

• Humming means that the fan is of poor design, with a lack of precision in the manufacturing process and a lack of noise-reducing material between parts.

• A poorly designed fan wobbles and has cheap bearings that are exposed. I lost a fan of my own because its cheap bearings were exposed. Dust got into the open bearings and locked it up.

Supporting a Heavy Fan

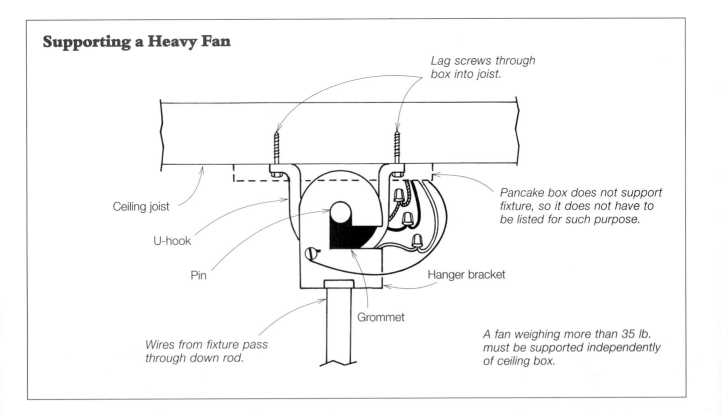

Lag screws through box into joist.

Ceiling joist

U-hook

Pin

Wires from fixture pass through down rod.

Grommet

Hanger bracket

Pancake box does not support fixture, so it does not have to be listed for such purpose.

A fan weighing more than 35 lb. must be supported independently of ceiling box.

• Make sure the fan comes with a long warranty—20 years or lifetime. If you don't get one, you'll wish you had later.

• The best fans keep up with technology. For example, some fans are equipped with remote controls (Hunter has several different remotes for its fans). Do some research and see what's new.

Fan design can add to the theme of the room. Don't be satisfied with the standard design if you want something different. I changed one of my typical ceiling fans to a replica of an 1886 oval, cast-iron design. Manufacturers make a wide variety of fan styles, including replicas of antiques, and even cartoon themes.

Hanging the Fan
When installing a ceiling fan, follow the manufacturer's instructions. For maximum comfort, position the fan as close as possible to the center of the room. The fan must be mounted securely via an

Seasonal Air Movement

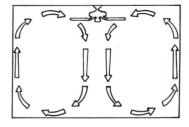

Summer cooling
Moving air cools room with fan in normal mode.

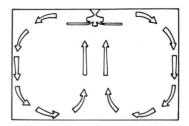

Winter warming
With fan in reverse mode, warm air is circulated downward.

approved plastic or metal box—the inspector may look for an approval sticker. If the box doesn't have a sticker indicating that it is rated for a ceiling-fan installation, don't use it. In the past, only metal boxes were allowed for ceiling-fan installations, but that's not the case today. You can now get a variety of nonmetallic boxes rated for ceiling fans. No matter what box you use, make sure it is securely attached to a stable ceiling joist. If the joist wobbles, so will the fan. If a joist is not close to where you need to install the fan, nail a 2x6 bridge from ceiling joist to ceiling joist and mount the fan to it. Approved fan-mount kits are available that will mount from joist to joist without having to go up to the attic (see the drawing below).

The weight of the ceiling fan is an important consideration. First, only fans that weigh 35 lb. or less, with or without accessories, can be supported directly from the ceiling box. Second, the box must be designed and listed especially for that purpose. Typical ceiling fixture boxes are not acceptable, even metal ones. Lastly, if the fan weighs more than 35 lb., it must be supported independently of a ceiling box. For such an installation, the fan manufacturer will supply a U-hook that screws into a wood beam in the ceiling. A rubber grommet fits through the center of the U-hook, and a hanger bracket, attached to the down rod, fits over the grommet pins on the outer edges of the grommet (see the top drawing on p. 189).

Reversible Fans

Most fans are reversible. In the normal mode, the fan blades push air down. With the flick of a switch, you can reverse the direction of the blades, allowing the blades to pull air upward (see the bottom drawing on p. 189). In warm weather the normal mode directs air downward, so the moving air cools the room by about 7°. It's possible to use your ceiling fan plus the

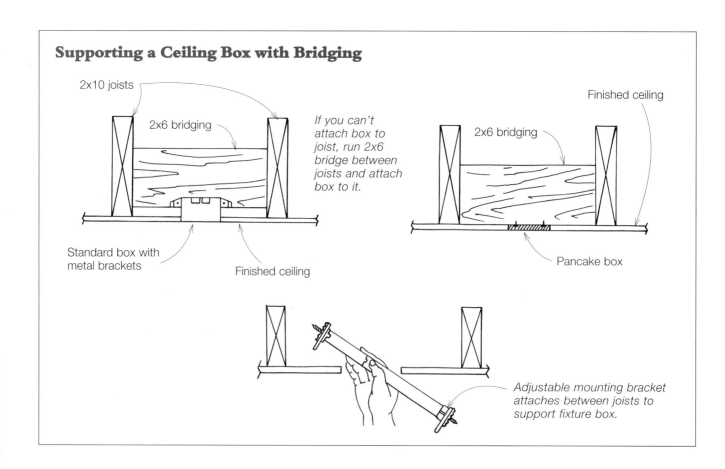

Supporting a Ceiling Box with Bridging

2x10 joists

2x6 bridging

If you can't attach box to joist, run 2x6 bridge between joists and attach box to it.

Standard box with metal brackets

Finished ceiling

Finished ceiling

2x6 bridging

Pancake box

Adjustable mounting bracket attaches between joists to support fixture box.

Two Ways to Wire a Fan/Light Combination

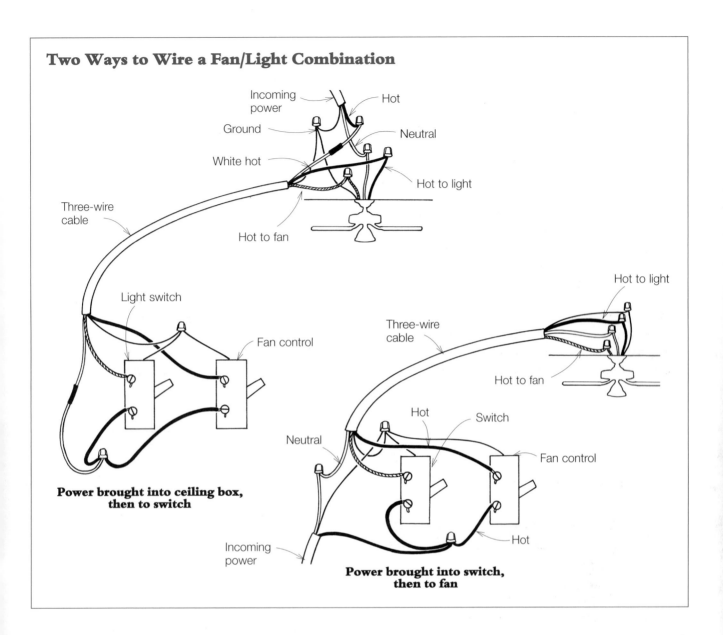

Incoming power

Hot

Ground

Neutral

White hot

Hot to light

Three-wire cable

Hot to fan

Hot to light

Light switch

Three-wire cable

Fan control

Hot to fan

Hot

Switch

Neutral

Fan control

Power brought into ceiling box, then to switch

Hot

Incoming power

Power brought into switch, then to fan

air conditioner in combination to cut cooling costs: set the thermostat at 79°, the fan at medium or high speed, and the room will feel like a comfortable 72°. In cool weather, reverse the fan's direction to pull air upward, which moves down the warm air trapped at the ceiling without creating a cool breeze.

Wiring a Fan is Easy

Wiring a ceiling fan is quite simple. Just follow the manufacturer's instructions. There will be a white neutral, a green or bare ground and two colored

wires—one for the light (if there is one) and the other for the fan. Remember, if you want to control both independently, the rough-in wire must have three conductors plus a ground—three-way switch wire. The wiring can be fed either to the fan first or to the switch first (see the drawing above). I prefer wiring first to the switch. Also, do not use a light control or dimmer as a variable-speed control for the fan motor. Instead, use a speed control that's designed for ceiling fans.

SMOKE DETECTORS

Smoke detectors save lives if installed properly. But sometimes they can be a nuisance. Smoke detectors are required by code to be installed in the hall outside bedrooms. This does not save you if the fire is in the bedroom. Ideally the detectors need to be both inside and outside the bedroom doors (this is required by the '94 Uniform Building Code).

Detectors must be hard-wired for new construction, and many municipalities also require a battery backup system within the unit. Most areas allow battery units for houses already built where the homeowner just wants to have extra protection. Code also requires that all hard-wired units alarm at the same time. To do this, all units must be wired together with three-conductor wiring—three-way switch cable, normally 14-3 (see the drawing below).

TRACK LIGHTING

Track lighting can provide accent lighting in various areas with only one cable bringing power. The problem with track lighting is that the lamps for one track could have a number of different wattages, which could add up quickly. It's very easy to go from a simple 60-watt circuit to over 1,000 watts by simply plugging in a few more lights in the track. This is not a significant problem with one short piece of track, but it is a problem with a long track or several tracks on the same circuit. I always wire track lighting on a 20-amp circuit unless the amount of track is small enough that I'm not concerned about additional wattage.

Track-lighting fixtures are available in a variety of colors, with black and white being the most common. Track sections are normally made of extruded aluminum and are available in 2-, 4- and 8-ft. lengths, with 12-gauge internal conductors. The tracks can be arranged in any order imaginable. The more common layouts are straight, square, L, angle, T and X. Some connectors have a feed option that allows the fitting to accept power directly into it. The tracks also come with myriad selection of connectors to accommodate every arrangement (see the drawings on the facing page).

Tracks are straight, which could be a problem if your walls or ceilings are not: You may see gaps between the track and the ceiling or wall. But don't be discouraged from using track lighting if your walls are not exactly straight. You can buy track units that hang from the ceiling, instead of running along the ceiling.

Wiring a Smoke Detector

Outgoing power (14-3 cable)

Ceiling box

Ground

Incoming power

Joist

Red wire triggers alarms on other units downstream.

Mounting strap

Neutral

Hot

Test button

White and black wires provide power to this unit and to others downstream.

Laying out Track Lights

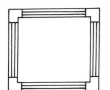

Square layout
Consists of four track sections and four L connectors with feed option.

Angle layout
Consists of two track sections, one flexible connector and one live end connector.

L layout
Consists of two track sections and one L connector with feed option.

X layout
Consists of one X connector with feed option and four track sections.

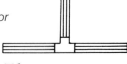

Straight layout
Consists of one track section and one live end connector.

T layout
Consists of one T connector with feed option and three track sections.

Track Connectors

Electrical end feeds

Live end connector and outlet-box cover for outlet-box feed.

Conduit adapter for surface wiring.

Live end connector feeds track from above with flexible conduit or BX.

T-bar end connector feeds track from above when T-bar is directly above track.

Electrical joiner feeds

T-bar adjustable connector feeds two track sections to make 90° turn or straight track run.

Outlet-box cover

Track joiners

Flexible connector can be adjusted to 90° to left, right or up or down and joins two track sections.

T-connector joins three track sections.

L-connector joins two track sections for 90° turn.

X-connector joins four track sections.

Two Ways to Bring Power to Track

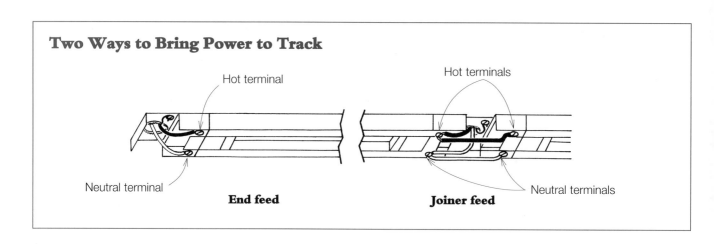

Hot terminal

Hot terminals

Neutral terminal

Neutral terminals

End feed

Joiner feed

Wiring Track Lights

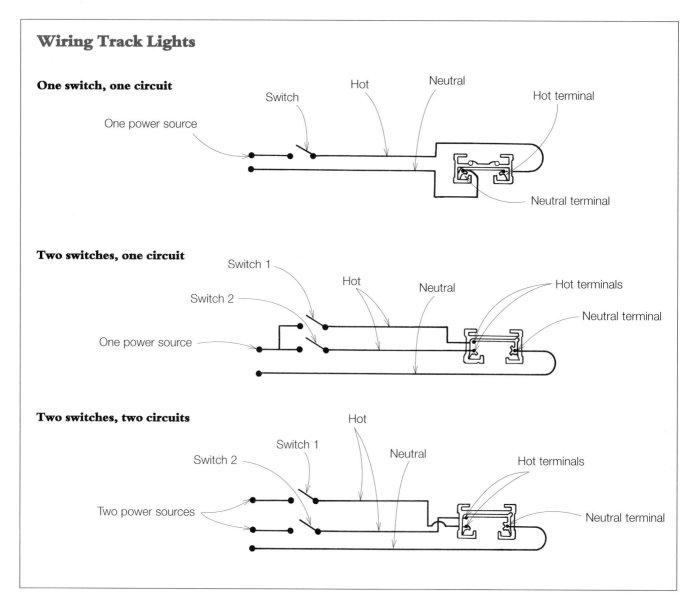

One switch, one circuit

One power source

Switch

Hot

Neutral

Hot terminal

Neutral terminal

Two switches, one circuit

Switch 1

Switch 2

Hot

Neutral

Hot terminals

Neutral terminal

One power source

Two switches, two circuits

Switch 2

Switch 1

Hot

Neutral

Hot terminals

Neutral terminal

Two power sources

Rough-in wiring for track lighting needs to be as exact as possible. Sometimes the rough-in wire can terminate in a standard outlet box. Most times the splice will be done in the end of the track or in the connector, so the cable must be brought through the finished wall at the *exact* spot where the track begins or ends (see the top drawing on the facing page). Make the hole through the finished wall no larger than the diameter of the cable. The power wires connect to screw terminals within the track, which has two or three 12-gauge wires running its length. The wires connect to the lights that clip into the track. Normally, lights are inserted into the track and turned 90° to lock them in (the number of lights installed depends on the length of the track and on your requirements).

Typically, track lighting can be low voltage or standard voltage, with one circuit or two circuits. A single circuit can have one switch-controlled hot wire and one neutral. A two-circuit track will have two hot internal wires sharing the same neutral. Each hot wire can share the same power source and be switched independently, or have two independent switched power sources (see the bottom drawings on the facing page). Lights from each circuit can be alternated within one track section or can be in totally different sections of track. If the track lighting is running off two independent circuits for the same track, be sure the two feed circuits are on opposite phases so that their neutral current can cancel. Because the track only has one neutral, these currents will add up if they're on the same phase and could possibly overload the neutral.

OUTDOOR LIGHT FIXTURES

The most common outdoor fixtures on a home are flood lights, entry lights, dusk-to-dawn lights and motion detectors. The wiring for these fixtures is pretty basic, but installation can be complicated.

Flood Lights

Flood lights are normally mounted outside at a high point on the house and come in wattages ranging from 75 to 1,500. Standard flood-lamp bulbs can be mounted up to three at a time, all facing different directions for maximum coverage. Don't mount these lights in an inaccessible location because you're going to have to change the bulbs at some point. If possible, mount the light above or adjacent to an upstairs window so that bulbs can be changed easily. Quartz lights can also be used for outside lighting. These bulbs give off a brilliant white light immediately in front of the fixture. But one quartz light cannot cover as wide an area as three floods facing different directions. These lights get hot, too, so don't mount them low enough where people can touch them.

For rough-in, simply bring the cable out of the building. Once the finished exterior is up, install a watertight box and the fixture. If you try to install a box ahead of time, you may not be able to match the exterior surface, and you will spend a lot of time trying to place a cover on the box that doesn't conform to the finished wall on the exterior.

Entry Lights

Entry lights can be troublesome if the building has beveled or irregular siding. If the fixture is placed flat against the siding, the light will slant along with the siding. If placed plumb to the wall, the light base will have a large gap underneath it because of the bevel of the siding. Here it's best to have the carpenter make a fancy base and install it for you. If that's not possible, grab a piece of siding and invert it on the installed siding to make a flat area. Once you've got a flat mounting surface, install a pancake box to mount the light. Roughing in simply involves sticking a cable through the exterior wall where the light is to be mounted. If the house has vinyl siding, you can get an adapter that will allow the fixture to seat on the exterior with a minimum of problems.

Dusk-to-Dawn Lights

Normally installed as security lights, dusk-to-dawn lights can be bought as an entire unit, or you can buy individual lights. Dusk-to-dawn lights normally have a plug-in photoelectric cell, or eye, sitting on top of the unit, which senses when it's dark and then turns on the light. These units are normally mounted in areas for parking or security. Packaged units normally use mercury vapor lights, while the individual sensor can turn on a light for any purpose—such as a sign to be lit after dark. These cells don't last forever—neither do the bulbs—so locate the fixtures where you can reach them easily when it's time to replace the bulb or cell.

Typical wattages are 70 to 175. Lower wattages may be cheaper, but they may not give off enough light. These lights are normally hard-wired on a 120-volt circuit. The disadvantage of such a fixture is that it could take several minutes for the unit to turn on. Separate photoelectric cells can be wired in series with any light to turn on after dark, but be sure the cell matches the wattages of the lights to be used, or you could burn it out.

Motion Detectors

Motion-sensing devices are very popular nowadays. Some units detect just motion, and other sophisticated models may also detect body heat using infrared rays. The units turn on house lights or floodlights as someone comes near the house. Their

loads are normally limited to around 500 watts. You can buy these as individual sensors, which can turn on lights that are already installed, or as complete sets with a light-attachment assembly.

Installing such a device can be problematic. First be sure the unit you buy can detect at the distance you want it to detect—30 ft. to 60 ft. is common (most units will have trouble detecting motion at farther distances). Mount the detector no higher than 12 ft. above the ground. The higher the unit is installed, the less ground surface that is covered, and the unit will be less responsive to motion on the ground. And locate the unit away from heat-producing sources, such as a heat pump, and away from reflective surfaces, such as windows and pools.

Also, it seems logical to face the unit into the direction of the object that will turn the unit on. But this is not what you want to do because these sensors—both infrared and motion—do not saturate the area. Instead, the sensor projects out like fingers

Locating a Motion Detector

Mount detector no higher than 12 ft. above ground and locate detector so that motion cuts lobes of detection at right angles. Locating unit straight into motion can create dead zones, where motion will not be detected.

110° field of view

Walkway

Dead zones

Detection lobes

60 ft.

Motion detector

Detection lobes extend straight into target area.

Not ideal

Walkway

110° field of view

Motion detector

60 ft.

Ideal

Detection lobes cut across target area at right angles.

Wiring a Typical Motion Detector

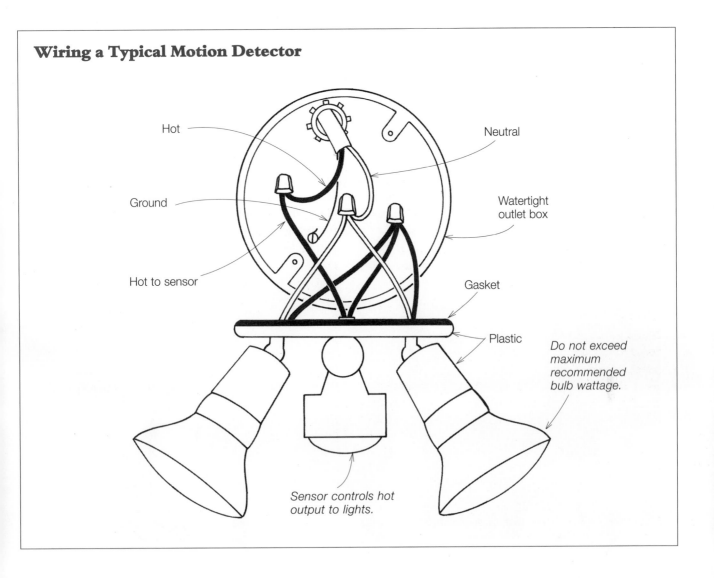

Hot

Neutral

Ground

Watertight
outlet box

Hot to sensor

Gasket

Plastic

*Do not exceed
maximum
recommended
bulb wattage.*

*Sensor controls hot
output to lights.*

(called lobes), so if you're facing the unit, you can actually walk to it—between the detection lobes—and not set it off. The directions on the unit tell you to install the unit at right angles to the area you want to track so that people will pass across the lobes of detection, not between them (see the drawing on the facing page).

The rough-in wiring only needs to be a cable sticking through the wall. Then attach a watertight box to the wall after the siding is up and pull the cable through. Make sure the box shape matches the

detector's shape—you want don't a round box for a square detector fixture. Once the box is up, simply follow the wiring instructions included with the unit (an example is shown above).

Avoid buying units that don't come with instructions (and there is at least one) or units that don't have sensitivity adjustments. A unit equipped with this allows you to adjust the range of detection anywhere from 20 ft. to 60 ft. You can also adjust the sensitivity of the unit to compensate for dogs, raccoons and other small animals.

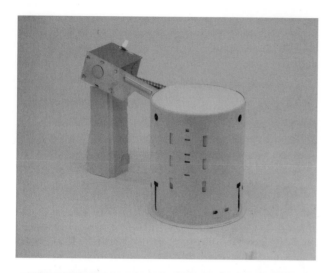

The recessed light fixture on the bottom is for new construction, and the one on top is for remodeling. Both are 6 in. (Photos courtesy of Direct Lighting)

Insulation and Recessed Fixtures

T Housings
Keep insulation 3 in. from fixture

ICT Housing
Okay to cover with insulation.

RECESSED LIGHTS

Recessed down-lighting can give a variety of illuminating effects throughout the house and can add light to an area without affecting other design criteria. They are commonly used in kitchens and closets, and for accent lighting or task lighting.

The housing is the main criteria of a recessed-light installation. You can buy housings for new construction or remodeling (see the photos above), and there are two basic types to choose from: T-

housings for noninsulated ceilings and ICT housings for insulated ceilings. T, or non-IC, housings must be kept at least 3 in. from any insulation in the ceiling or walls. All housings are equipped with a thermal protector that will shut the light off if insulation material is installed too close to the housing or if the lamp wattage is exceeded.

IC fixtures sometimes have an ICT housing (standard round), UICT housing (double-wall rectangular) or an ICAT housing (airtight). The ICs are the preferred

Wiring a Recessed Light

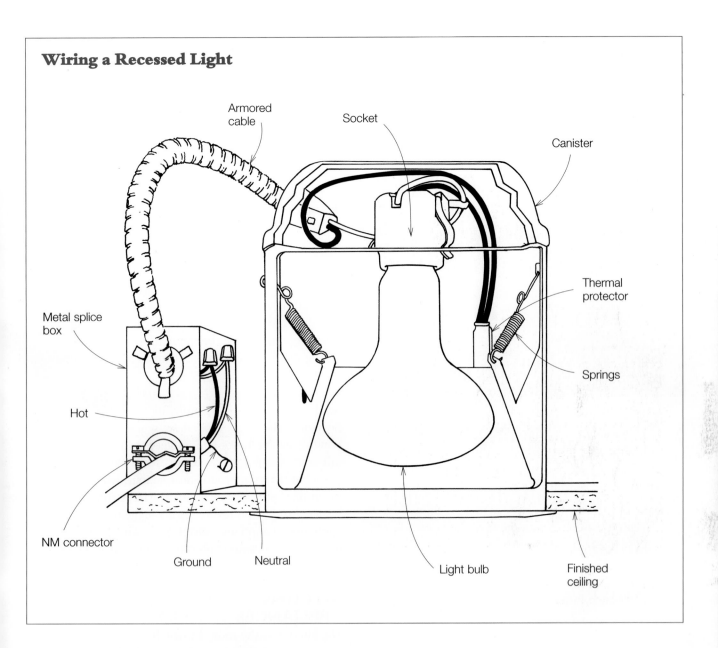

Armored cable

Socket

Canister

Thermal protector

Springs

Metal splice box

Hot

NM connector

Ground

Neutral

Light bulb

Finished ceiling

housings because they may be completely covered with insulation (see the drawing on the facing page). But such housings may have wattage listings a little lower than those of T fixtures. Housings are also available for special applications: sloped ceilings, closet lights, deep baffle housings designed for superior visual cutoff and housings for high wattage lamps (up to 200 watts). For remodel jobs, verify that there is enough room in the ceiling height for the fixture. If height is a problem, shallow models are available. Use only trim that is compatible with the particular fixtures you're installing—look for warnings by the manufacturer.

Wiring is normally done via a metal splice box that is attached to the housing frame (see the drawing above). Rough-in includes mounting the housing but not the trim plate. When mounting the housing, be sure to allow for the finished ceiling (in new construction).

Chapter 12

WIRING APPLIANCES

Our homes are becoming more and more appliance-oriented. It seems the more appliances we have, the more affluent we see ourselves as—and the trend doesn't seem to be slowing. With the increasing number of appliances being included in a house, wiring them is taking more of an electrician's time on the job site.

Appliances are not the simple hookups they used to be. Gone are the good-old days, when you simply bought the appliance and plugged it in. Today it's different. While some appliances still have a cord-and-plug hookup, others must be hard-wired, meaning that the branch-circuit wire is connected straight to the appliance. Also, voltage from appliance to appliance varies: Some are 120 volts, others 240 volts, and still others are both 120 and 240. And they all wire differently.

KITCHEN APPLIANCES

The kitchen is the room in which you probably have the most electric appliances—dishwasher, refrigerator, garbage disposal, microwave, stove and a slew of others. Some of these appliances are plug and cord—some of which require a dedicated receptacle—but many need to be hard-wired on a dedicated circuit. In this section I'll give you some common-sense approaches to wiring kitchen appliances, and I'll throw in a few tips on choosing appliances and their accessories.

Dishwasher

Residential dishwashers are normally 120 volts and require a dedicated 20-amp branch circuit using 12-gauge wire. The cable is brought through the lower part of the wall to the back right or back left of the appliance (bring in enough slack so that the dishwasher can be pulled out and serviced). The wire is brought into a metal splice box located in the front of the dishwasher (normally in the bottom right corner). The box is accessed by removing the kick panel at the bottom front of the dishwasher.

The first step is to install an NM connector, or strain relief, on the splice box, then you can bring in the wire. Wire connections are simple: hot to hot (black to black), neutral to neutral (white to white) and ground to box (see the top drawing on p. 202). If the house wiring is old and has no ground, you'll need to run an equipment-grounding wire to the main panel or install a new cable. The dishwasher has a switch that can disconnect all electrical power to itself, so it does not require a separate service disconnect (see the sidebar below).

Cutting Power Off for Maintenance

When servicing an appliance, you obviously need to cut off power to the appliance to work safely. For most plug-in appliances, where the cord and plug are accessible, this is easy— just pull the plug. However, a hard-wired appliance is a different story. You would think that you could simply turn the circuit breaker off at the main panel, but someone could come along and turn the breaker back on, not knowing that another person is working on an appliance on that circuit.

If the hard-wired appliance is out of sight of the main panel, a service disconnect—a cutoff switch located adjacent to the appliance—is required. If a disconnect cannot be installed, or if the homeowner does not want one because of aesthetics, a lockout must be used in lieu of the disconnect. A lockout is a device put on a circuit breaker

that allows you to lock the breaker in the OFF position while you work on the appliance so that the breaker cannot be turned back on (see the drawing below). Some lockouts allow you to padlock the breaker off, others lock the breaker with an integral key and lock, and yet

others are simply mechanical. A lockout is required for any appliance that does not have a single switch to turn off all its power, that is out of sight of the main panel and has no means of disconnection at or near the appliance. The panel lock does not qualify.

Lockouts

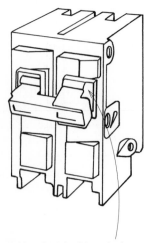

Handle-blocking device

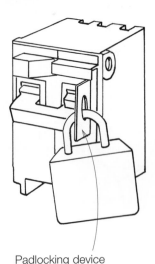

Padlocking device

Wiring a Dishwasher

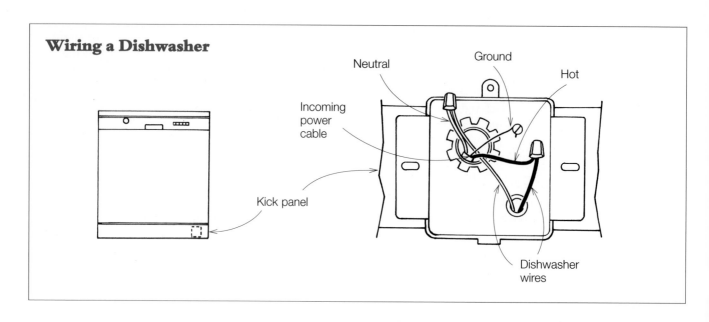

Neutral

Ground

Hot

Incoming power cable

Kick panel

Dishwasher wires

Hard-Wired Garbage Disposal

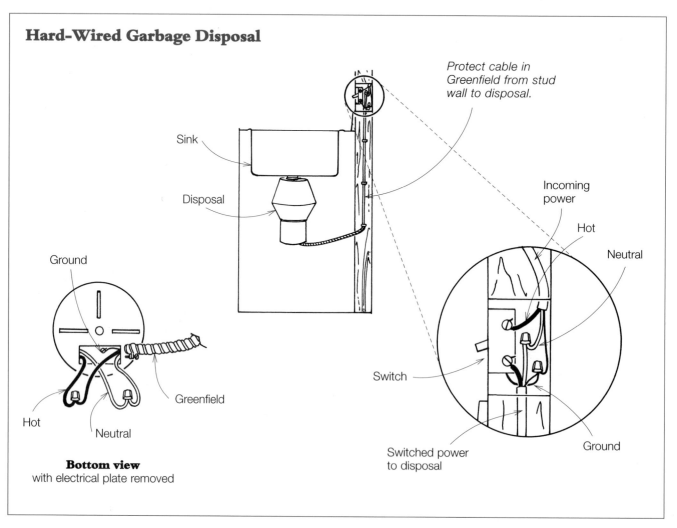

Protect cable in Greenfield from stud wall to disposal.

Sink

Disposal

Incoming power

Hot

Neutral

Ground

Hot

Neutral

Greenfield

Switch

Switched power to disposal

Ground

Bottom view
with electrical plate removed

If you prefer to attach a plug and cord to the dishwasher, not hard-wire it, keep the length of the cord to between 3 ft. and 4 ft. The receptacle must be accessible and adjacent to the appliance—it's normally installed right under the kitchen sink.

Hot-Water Dispenser

Instant hot-water dispensers are gaining in popularity and provide ⅓ to ½ gal. of 190° water for coffee, tea, hot chocolate and soup. These normally install into the sprayer hole in the kitchen sink, but you can punch a hole anywhere. A typical unit is connected by a cord and plug (3-ft., grounded, 16/3), with a wattage of between 500 and 750. You simply have to wire in a receptacle under the kitchen sink. The branch-circuit wiring for one of these units is normally 14 or 12-2 w/g. And because a hot-water dispenser only pulls about 4 to 7 amps, it doesn't require its own dedicated branch circuit. However, it is common for one 20-amp branch circuit to be dedicated if both a garbage disposal and hot-water dispenser are on the same circuit. A hot-water dispenser has a plug and cord, which can be unplugged for servicing, so it does not require a separate disconnect switch. However, some interpreters of code say any fixed appliance cannot be plug connected.

Garbage Disposal

Almost all garbage disposals available today are called continuous feed. A continuous-feed unit will require a switch to be installed, normally on the kitchen wall adjacent to the sink. As long as the switch is on, you can feed kitchen scraps continuously into the unit. A garbage disposal also requires a disconnect for maintenance if it is hard-wired, and the wall switch can serve as the disconnect—nothing else needs to be wired in.

Garbage disposals are wired from the bottom, either through a plug and cord or hard-wired. If the unit is hard-wired, the cable will come right out of the drywall under the kitchen sink or straight up from the crawlspace through the bottom of the cabinets—be sure to keep the wire out of the way

of the plumbing (it should be run within Greenfield). Voltage is normally 120, and most garbage disposals are not required to have their own dedicated 15- or 20-amp branch circuit. The electrical connection is made under the electrical cover, or plate, on the bottom of the disposal itself. The splices are made with wire nuts: black to black, white to white and the ground to the frame screw on the unit (see the bottom drawing on the facing page).

To control power to the disposal, install a switch on the wall near the kitchen sink. If the disposal is plug and cord and if an instant hot-water dispenser is also at this location, they both can be plugged into the same receptacle. For this setup, however, I would recommend that the receptacle be wired on its own dedicated branch circuit.

Electric Stoves

Electric stoves are 120 *and* 240 volts. The burners or bake unit require 240 volts, but the timer, clock and buzzer use 120 volts. The NEC gives specific guidelines for wiring stoves, including choosing the right-size conductors.

For a wall-mount oven and a drop-in range or cooktop, the NEC requires you to use the nameplate rating to determine the correct-size conductor to use. However, the NEC allows the typical slide-in stove to be derated. That is, if the nameplate rating is not over 12 kilowatts, the rough-in wiring can be designed for 8 kilowatts. The NEC allows this derating because it is unlikely that all elements will be on high, or full power, for an extended length of time. However, I feel the NEC is wrong because I've had to replace many 40-amp slide-in stove circuits with 50-amp circuits because excessive current kept kicking out the 40-amp breaker. According to code, a standard electric stove is only going to pull 30 to 33 amps max. However, if all burners are going, along with the bake unit—which could happen while cooking a holiday dinner, when a lot of family and friends will be over to enjoy a feast—they can easily pull 40 to 50 amps. I think it's

Wiring Stove Receptacles

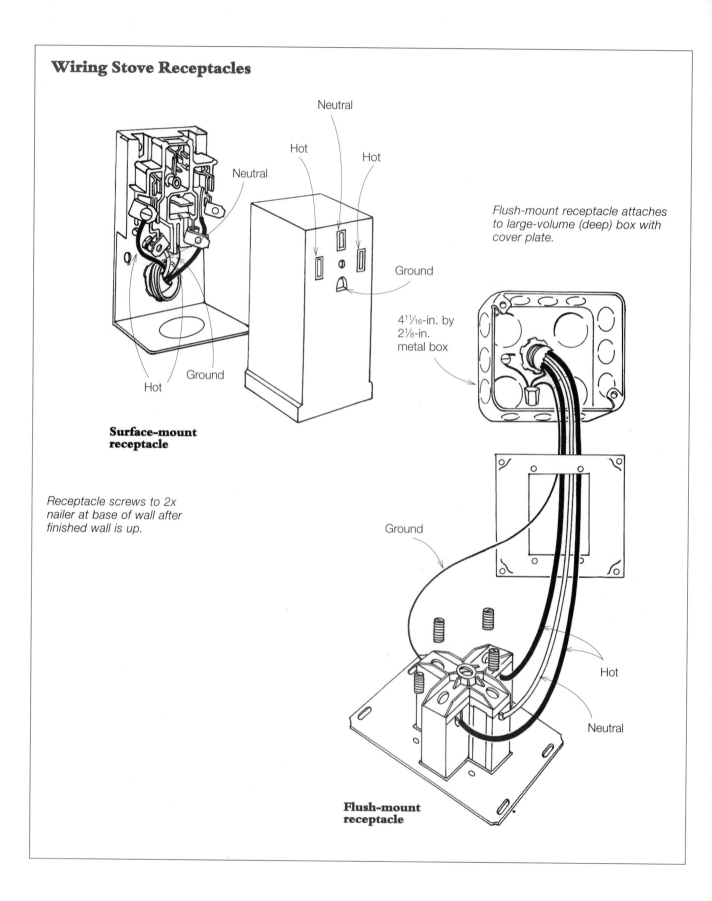

Neutral

Hot

Hot

Neutral

Ground

Hot

Ground

Surface-mount receptacle

Flush-mount receptacle attaches to large-volume (deep) box with cover plate.

$4^{11}/_{16}$-in. by $2^{1}/_{8}$-in. metal box

Receptacle screws to 2x nailer at base of wall after finished wall is up.

Ground

Hot

Neutral

Flush-mount receptacle

Wiring Drop-In Cooktop and Oven

Cooktop whip

Oven whip

Hot

Master splice box
in cabinet in wall
or flush to wall

Incoming power (6-3
w/g NM cable)

Neutral

Box grounding screw
(whip ground wires may
not be present if AC is
used)

*Conductors in both whips are
encased in Greenfield (AC).*

best to design the stove circuit for 50 amps (use 6-gauge copper) instead of allowing for the NEC's derating standards.

A standard slide-in stove will plug into a 50-amp female receptacle. The receptacle should be mounted low so that the bottom panel or drawer of the stove can be removed to disconnect the unit. This way you won't have to install a disconnect switch (see NEC 422.22b). The outlet box can be surface-mounted or recessed. I normally attach a 2x nailer between studs, bring the cable through the nailer, and screw the receptacle to the nailer after the finished wall is up. I think it's the easiest way to go. A more complicated installation is to use a recessed box and install a flush-mount receptacle. You'll have to choose a $4^{11}/_{16}$-in. by $2^1/_8$-in.-deep square box (42 cu. in.) because of the large cable required (see the drawing on the facing page), and put a large cover plate over that for the receptacle to mount to.

Because a stove pulls both 120 and 240 volts, the neutral must be insulated (this applies to hard-wired units, as well). As of the 1996 code, three-prong receptacles, both surface and recessed, are no longer allowed. The ground and the neutral cannot be connected, meaning the neutral must be an insulated conductor separate from the ground. Now you need a male plug and female receptacle with four conductors: two insulated hots, one insulated neutral and one ground.

If you've got a drop-in cooktop—one that's mounted in the countertop—and the oven is wall-mounted, you must install a lockout at the main panel and wire both through a master splice box. If the electric range or oven is a drop-in type, there should be a pigtail of wires, called a whip, within flexible conduit, hanging from the unit. These will connect to the incoming power cable in a large splice box supplied by you (see the drawing above). Simply bring the four-conductor cable into the box to splice to the range's wires. If the whip is not installed already, do not be tempted to use a plug and cord on a drop-in cooktop—some models even come with labels warning against their use. Instead, make your own whip, using THHN wires (12 or 10 gauge, depending on the load) and run them through Greenfield. Or you can use a leftover piece of cable.

Range hood and fan Range hoods can be almost any size, from 30 in. up to 72 in. long. You can get a standard economy hood (I'd avoid this type because it doesn't pull enough air), to the old-style hammered-metal canopy hoods. Also available are high-tech models, with features such as blower speed control; heat-sensitive components that turn the blower automatically to high speed when excess heat is sensed; warming lamps to keep food hot; high-heat alarms; and dual lighting.

The whole purpose of a range hood is to move air. The blower motors on economy models are very small and move little air (around 140 cfm or less). However, units with large dual blowers (up to 1,200 cfm) are available from some manufacturers. If you don't want a noisy fan (it is impossible to have the power you really need and keep the noise to a whisper), mount the blower on the outside of the

house. Broan, for example, has a roof- or wall-mounted 900-cfm blower available for its Rangemaster hood. This type of blower assist looks like an inverted aluminum bowl mounted to the side of a house.

Most fans are 120 volts with the rough wiring connecting to a box within the unit. Because it pulls little current, the fan can be wired into the kitchen light circuit. The problem with wiring the stove fan is that, during the rough-in stage, the wire has to be brought out of a bare stud wall when the fan, or anything else in the kitchen, isn't there yet. Once you've got the rough wiring run to the correct spot, staple it in place on the stud, with about 18 in. of the cable sticking out beyond the stud wall (see the drawing below). Then comes the drywall and the kitchen cabinets. After that, the hood and fan unit will mount to the bottom of the cabinets. The rough-in wiring must be brought out exactly under the proposed bottom of the overhead kitchen cabinets, but not too far down as to be exposed once the fan unit goes in. If the cable comes through too high, you can cut the drywall a bit more to get the cable

Rough-in for a Range Hood

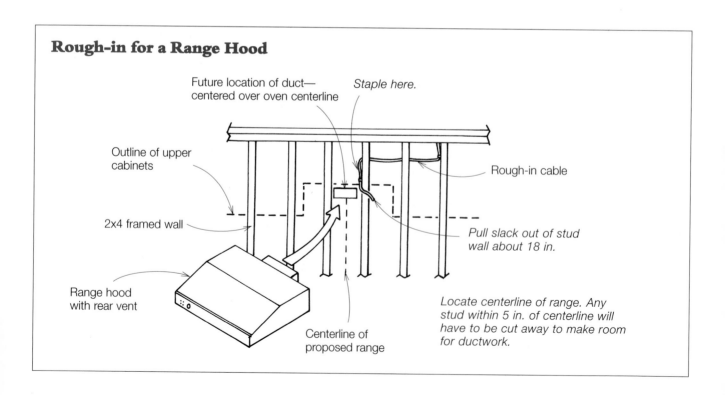

Future location of duct—centered over oven centerline

Staple here.

Outline of upper cabinets

Rough-in cable

2x4 framed wall

Pull slack out of stud wall about 18 in.

Range hood with rear vent

Centerline of proposed range

Locate centerline of range. Any stud within 5 in. of centerline will have to be cut away to make room for ductwork.

down to its correct location, as long as the hole you cut will be covered by the fan housing and the overhead kitchen cabinets. If this isn't the case, you'll have some patching to do, which could get sloppy. The wiring is pretty simple: black to black, white to white, plus the ground.

Standard installation height for fans is about 54 in. to 60 in. from the floor to the bottom of the unit. Add another 5½ in. to 9 in. to that for the height of the unit itself. Try to bring the cable out of the wall about 7½ in. to the right of the centerline of the duct. Be sure the drywallers do not move the cable location or cover it. If you're responsible for running the ductwork, make sure there is no stud dead center of the proposed location—you normally need at least 5 in. of clearance on either side of dead center.

Running the ductwork Duct size can be round (6 to 9 in. in diameter) or rectangular (3¼ x 10 in. is standard). When running the ductwork for long runs, you don't have to run the rectangular duct all the way to the outside unless you just want to: It's possible to make a transition to round duct with adapters made for that purpose. It's nice to have the range hood in hand so that you know what size duct, if any, you need to design for. If there is no room for ductwork, fans equipped with carbon filters are available, which clean the air and then blow it back into the room. These, however, gum up quickly, especially if you do a lot of greasy cooking. Use an outside-vent fan if at all possible. Make sure the air outlet vent is a minimum 3 ft. to openings (windows and doors). I also recommend buying splash plates, which make cleaning the wall behind the range easy.

There are two alternatives to running the duct straight through the wall to the outside—neither of them is good. You can take the duct straight up into the overhead cabinetry and into the attic. But this wastes cabinet space, and you may have to cut it to pieces. Or you can make a bend inside the wall cavity and from there go straight up. To do this means completely cutting away the double top plate in that wall cavity, which is never a good idea. Keep it simple: If at all possible always go straight out with the duct.

One last piece of knowledge: Don't do what I did when I once attached the fan hood to the cabinet bottom. I used too long of a screw (four of them) and went all the way through the cabinet bottom and into a fancy menu holder and some Tupperware. This is one of those errors you don't make twice—use screws short enough that they don't come through the wood.

Trash Compactor

Some built-in compactors can be hard-wired. But these units normally come with plug and cord attached. Just be sure to have the receptacle accessible. If the unit is hard wired, and if there is no switch to turn off the unit's power, you will have to install a service disconnect switch or a lockout at the breaker. Read the instructions to verify if the manufacturer wants a dedicated branch circuit.

Microwave Oven

Microwave ovens are common in almost every household. However, so are the troubleshooting service calls because of them. Many times the breaker will trip when other appliances are running at the same time. Because of this, many manufacturers require 15- or 20-amp dedicated circuits. The instructions that come with the oven will indicate if your unit does or does not.

BASEBOARD HEATER

Electric baseboard heaters are a fast and economical method of getting heat where you need it. They are normally installed below windows and on adjacent walls. Electric baseboard heaters can be a fast and easy installation, as long as you know what to do and the problems to look for. The first decision to affect the wiring is deciding whether the thermostats are going to be inside the heater or on the wall. Inside the heater makes for an extremely fast and economical installation. However, some people like the wall thermostats better because they don't like bending over to adjust the thermostat.

Though 99.9% of all baseboard heaters are rated to use 240 volts, 120-volt models are available—but I don't recommend them unless there is only 120 volts available. Chances are high that the heater could accidentally be installed on a 240-volt circuit, resulting in a fire—this has happened. In addition, a 240-volt heater draws *half* the current of a 120-volt heater, which is more efficient and uses less expensive wiring. It will be assumed that all heaters in this discussion will be 240 volts.

The wattage of the baseboard heater will determine the heat you desire. Most panels are 250 watts per ft., or about 1 amp per ft. To figure the exact amperage, divide the wattage by the voltage. For example, say you're putting a 10-ft. unit (2,500 watts) under a picture window in a dining room. By dividing 2,500 by 240, you see that the amperage is just above 10.

Several small baseboard heaters can be fed from the same cable coming from the main panel as long as the current doesn't exceed 80% of the cable's rating: 12-gauge cable cannot take more than 16 amps continuous current; 10 gauge cannot take more than 24 amps; and 14 gauge cannot take more than 12 amps. (I do not recommend 14-gauge cable for baseboard heaters because it's too small.)

Bring Power into One End

Baseboard heater

Leave one end alone

Untwist wires on one section end and wire incoming power and thermostat there.

When power is first applied to a new heater, it will smoke, smell and burn off the factory coating on the elements. This is normal and will stop after a few minutes. It's best to open the doors and windows and let the room air out as the heater is broken in. In addition, heaters unused for an extended time will collect dust and cobwebs on the elements. It's best to remove the front cover of the baseboard (it will snap off) and vacuum the element before turning on the unit.

If you need a receptacle in the wall over a baseboard heater, use a receptacle designed to be installed in the end or center of the baseboard: Never install a receptacle over a heater. It's against code and could burn any cord that touches the heater or hangs very close to it. Bring in a separate line to power the 120-volt receptacle—it cannot tap off the 240 volts powering the heater.

Baseboard heaters normally come with two end sections: Either end can be used for wiring the thermostat. Once one of the end sections is opened, you will notice two wires twisted together with a wire nut, which are the two hot wires for the heater. Untwist them, remove the wire nut and leave the other end section alone, with its wires spliced (see the drawing at left).

Wiring a Wall-Mount Thermostat

There are two types of wall-mounted thermostats: single pole and double pole. Both are rated for 240 volts. The single-pole unit is less expensive than a double pole, but it only switches one leg of the 240-volt circuit. Switching the heater down to its minimum setting of 40° with the thermostat control will normally stop the heater from working, but because one leg of the circuit is still hot, it could shock or electrocute an unwary electrician unless he or she throws the breaker first.

The double-pole unit has an off switch that allows you to open both legs of the circuit. This is the better unit, and the only one now allowed by the NEC to be installed in the wall. In addition, most homeowners prefer this type of thermostat because of the piece of mind the off position gives them. For those who currently have a single-pole thermostat

Wiring Wall-Mount Thermostats

Single-pole thermostat is currently noncode and should be replaced with double-pole unit.

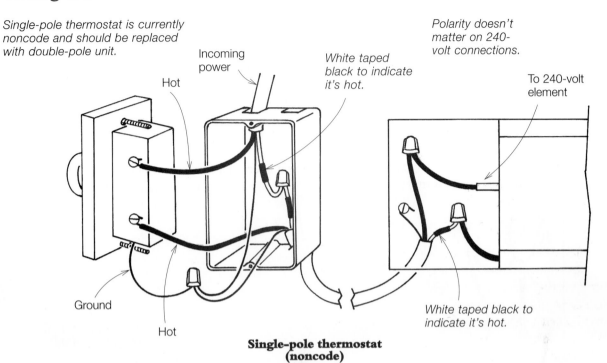

Incoming power

Hot

White taped black to indicate it's hot.

Polarity doesn't matter on 240-volt connections.

To 240-volt element

Ground

Hot

White taped black to indicate it's hot.

Single-pole thermostat (noncode)

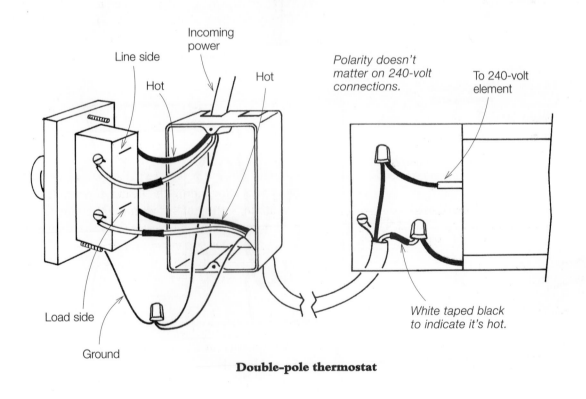

Incoming power

Line side

Hot

Hot

Polarity doesn't matter on 240-volt connections.

To 240-volt element

Load side

Ground

White taped black to indicate it's hot.

Double-pole thermostat

and want to update to the new codes, simply replace the single pole with a double-pole unit. Here's how: First remove power from the circuit. (If there is only one wire, a single-pole to double-pole conversion will be impossible.) Verify that power is off with a voltmeter, then remove the thermostat and take off the wire nut from the white wires. Install the double-pole unit. Wire incoming power to one side (line side), and outgoing to the load side (see the drawing on p. 209).

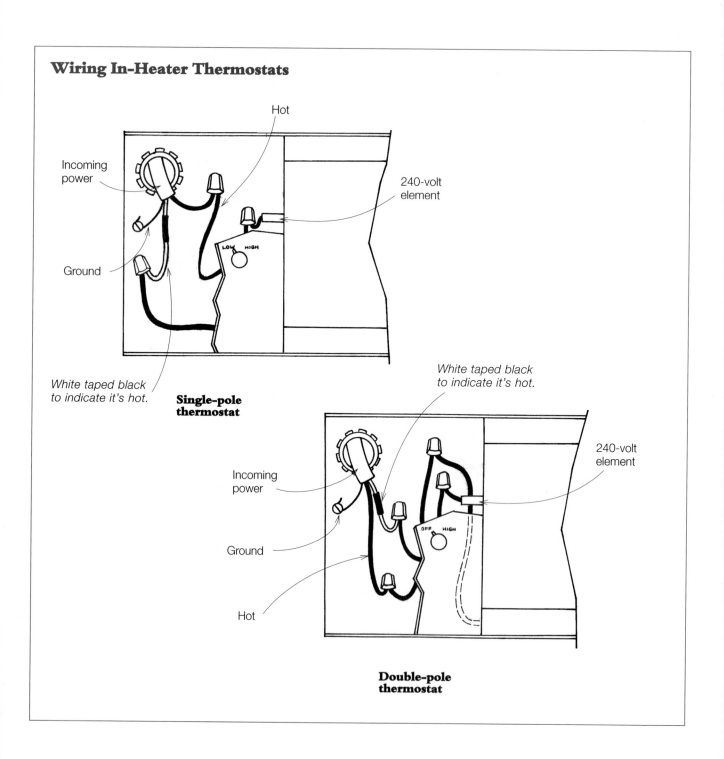

Wiring In-Heater Thermostats

Hot

Incoming power

240-volt element

Ground

LOW HIGH

White taped black to indicate it's hot.

Single-pole thermostat

White taped black to indicate it's hot.

240-volt element

Incoming power

OFF HIGH

Ground

Hot

Double-pole thermostat

For new installations, instructions with the wall thermostat will tell you the mounting height of the standard rectangular rough-in box. Which wall you install the thermostat on will be up to you and the instructions (don't install it below a window or above a heater). One thermostat can control several heaters within a room, as long as they are in sight of the thermostat and the current rating of the thermostat isn't exceeded; the current rating is listed in the instructions (it's normally around 20 amps).

The wiring must run to the wall thermostat first and then to the heater. Thermostats and their wiring take up a lot of space. Be sure the rough-in electrical box is deep—minimum 3¼ in. deep, or 20 cu. in. A single-pole thermostat will splice one leg directly through, with the other being connected through the thermostat. The thermostat will indicate which of its screws the incoming power is to connect to and which the load goes to. There will be a ground connection as well. The ground screw on the thermostat can only have one wire underneath it, so you'll need to pigtail a ground wire off a grounding splice. Double-pole thermostats take both legs of the power cable through its switches.

A typical mistake that will destroy both a single- and double-pole thermostat is wiring the incoming power into a direct short through the thermostat without having the load on line. To avoid this mistake, wire the thermostat in series with the heater: a hot wire to one side of the thermostat, and the load (heater) on the other. If you wire hot to one side of the thermostat and the other hot wire to the same side of the thermostat switch, it will short and throw the breaker as soon as the thermostat closes its contacts.

Wiring an In-Heater Thermostat

An in-heater thermostat is more economical than a wall-mounted thermostat. A wall thermostat turns all units within a room on at one time; instead of placing heat where it's needed in an area of the room, all units will click on at the same time, which wastes energy. Heaters with integral thermostats turn themselves on and off independently, resulting in less overall power consumption.

Installing the thermostat in the heater is easier than installing the thermostat in the wall. Simply pull off

the cover plate on one end of the unit and put the thermostat on. The rough-in wiring can be run straight to the heater's location, through the wall and the knockout in the heater. Double-pole in-heater thermostats are available, but most are single pole (though a wall thermostat is no longer allowed to be a single pole, in-heater thermostats can be). Therefore, one incoming hot wire will splice straight to the heater, and the other will splice to the thermostat; the other end of the thermostat will connect to the heating element (see the drawing on the facing page). This places the thermostat in series with one leg of the incoming power. Again, if you splice the incoming power across the thermostat, a direct short will be created when the thermostat engages, and the thermostat will be ruined.

ELECTRIC WALL HEATERS

These come with and without fan and in both 120- and 240-volt versions. Always buy the fan-operated unit because it will distribute the heat around the room better. This is especially important if plumbing is in the room because heat will rise, and you'll wind up heating only the top of the room, allowing the plumbing to freeze. Always buy the 240-volt model: It draws half the current of the 120-volt unit. You are also allowed to wire more than one heater in a branch circuit, as long as the wiring isn't overloaded.

Many wall heaters do not come with thermostats. The ones that have thermostats install much faster because you don't have to install another outlet box and wire a separate thermostat. Whichever you prefer, be sure to check the label on the package to make sure you get the type you want. Electric wall heaters are fairly inexpensive and install quickly— one cable, one hookup, and you're done. A small, single-wattage heater is ideal for a small room, like a bathroom. Large, variable-wattage units with fans are better to heat large rooms or even a whole house— heating the house with this type of heater is economical only if the areas are open. The large unit is about 6 ft. tall and can fit between studs or can be mounted on a wall. It can also heat an adjacent room if you install a length of duct from the heater to that room.

UTILITY-ROOM APPLIANCES

Wiring the appliances in a utility room can be tricky. Many of the appliances are covered specifically by the NEC, and each situation is different. In general, the appliances in this room are large—water heater, heat pump, washer and dryer—and may use both legs of a 240-volt circuit or just one. Wiring appliances this large is not a job for an amateur.

Water Heater

The standard electric water heater is a 50-gal., 240-volt, 4,500-watt, storage-type appliance fused at 30 amps. It is hard-wired with 10-2 w/g NM cable

Wiring a Water Heater

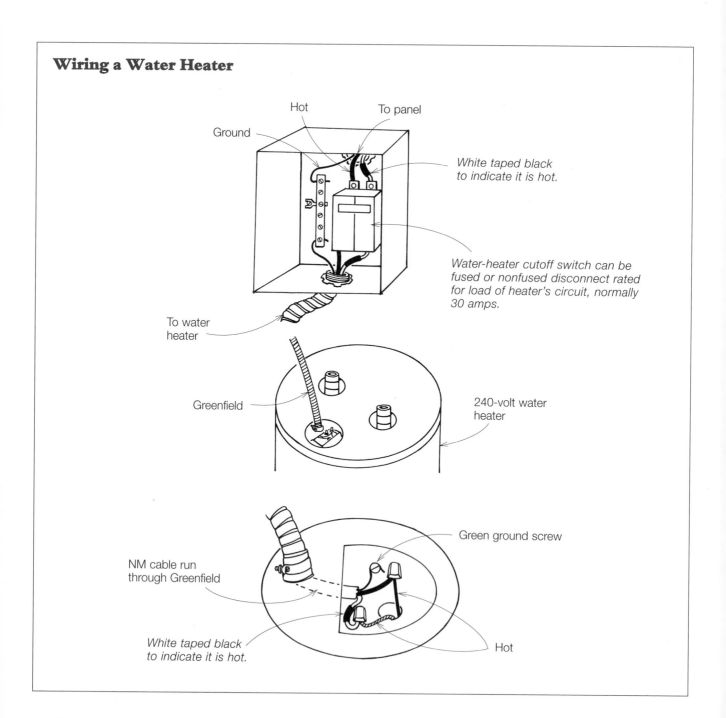

Hot

To panel

Ground

White taped black to indicate it is hot.

Water-heater cutoff switch can be fused or nonfused disconnect rated for load of heater's circuit, normally 30 amps.

To water heater

Greenfield

240-volt water heater

Green ground screw

NM cable run through Greenfield

White taped black to indicate it is hot.

Hot

run from the main panel to a disconnect, or cutoff, switch, to the heater. A cutoff switch (or lockout) is required if the heater is out of sight of the main panel, and the heater should be as close as possible to the cutoff switch. If the heater is in a habitable area, the NM cable from the switch to the heater must be enclosed in metal conduit (flexible conduit or Greenfield is best for this situation).

Bring the NM cable out of the wall and into the back of the cutoff switch. From the cutoff switch, swing the conduit—with cable enclosed—over to the heater. In the heater's splice box, connect the two incoming insulated hot wires to the two hot wires of the heater—no polarity is required, so don't worry about matching the colors of the two hot wires. Because the heater is on a 240-volt circuit, both white and black wires are hot, so cover the white with black tape at the heater, at the cutoff and at the main panel to indicate that it is hot, not neutral (see the drawing on the facing page).

Be sure to ground the water heater via the ground screw provided at the splice box. The grounding wire plays an extremely important part in safety around the heater. I've had many a service call on ungrounded water heaters. On one, the homeowner complained of getting shocked off the copper lines as he was trying to repair a split where the pipes had frozen. Sometimes he would receive a shock, and other times he wouldn't. Tracing the current fault back to the water heater, I noticed that the water heater was ungrounded. I pulled off the bottom cover and saw where the element wire (with screw still attached) had pulled away from the element and was touching the metal case of the heater. All was a mass of corrosion. Every time the upper thermostat would apply power to the bottom element, the electricity was being shorted over to the heater jacket and then into the copper lines. This is an extremely common problem with ungrounded water heaters. An ungrounded water heater should be grounded immediately.

If you want to wire a 240-volt water heater with 12-gauge cable, the heater will need a 3,500-watt element or smaller. There is nothing wrong with using a smaller than normal wattage; it simply takes longer to heat the water. However, because the water heater has a label on its jacket that lists the specs of the unit, there is a possible safety risk. When the plumber or do-it-yourselfer replaces a blown element, he or she may accidentally replace it with a 4,500-watt element, instead of the smaller 3,500-watt element. The large element will overstress the 12-gauge cable with excessive current and could start a fire. Therefore, be sure to write on the water heater—in large letters—that the 4,500-watt element has been replaced with a 3,500-watt one. Personally, I don't think it's worth the risk.

The gauge of the wire is determined by the current draw of the appliance. For a typical 4,500-watt/240-volt heater, a clamp-on ammeter will read around 18.75 amps. However, code dictates that the wiring to storage water heaters of 120 gal. or less be designed for no less than 125% of the load (NEC, 422.4a, ex. 2 and 422.14b). This means you must design the wiring for a 5,625-watt (23-amp) heater. Since 12-gauge cable can only carry 20 amps of current, you must use 10-gauge in this case. And because different-wattage elements can be easily swapped (5,500-watt elements can be easily installed in any size water heater and will pull 29 amps of design current, 23 amps actual), I recommend always using 10-gauge cable for electric water heaters. No other taps are allowed on the water-heater branch circuit to service other appliances. The maximum overcurrent protection allowed is 150% of the rated current—150% of 18.75 amps is 28 amps (NEC, 28e), but you can use the next-size breaker, which is 30 amps (NEC, 240.6a). The water heater must be on a dedicated circuit.

Electric Dryer

Dryers are the most miswired of all the appliances because they are both 120 and 240 volts. In addition, a residential (not mobile home) dryer may have an internal strap connecting the neutral to the frame. This strap must now be removed at the factory or in the field because the NEC requires a dryer to have its neutral separate from the frame ground.

Receptacle Wiring for Electric Dryer

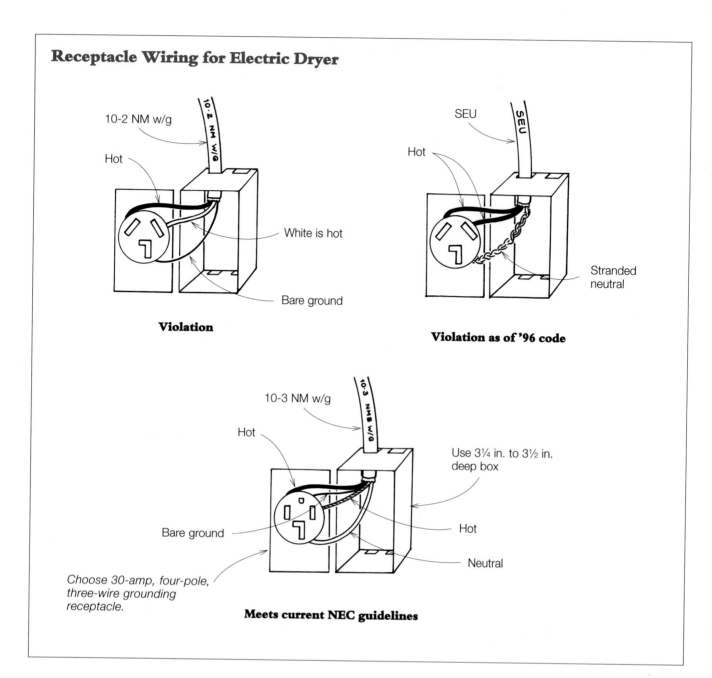

10-2 NM w/g

Hot

White is hot

Bare ground

Violation

SEU

Hot

Stranded neutral

Violation as of '96 code

10-3 NM w/g

Hot

Use 3¼ in. to 3½ in. deep box

Bare ground

Hot

Neutral

Choose 30-amp, four-pole, three-wire grounding receptacle.

Meets current NEC guidelines

I've seen hundreds of dryers wired with only two insulated conductors, such as 10-2 NM w/g. Though it's true they still have a good safety record, it is now a code violation. The heating element is 240 volts and is fed by the two insulated hot legs of the branch-circuit feeder wire. The timer, the buzzer and other parts of the dryer are 120 volts, and the return feed is through the neutral conductor. *Do not attach the bare ground wire to the center neutral connection: It's extremely dangerous and is a code violation.* When done, it places current flow in parallel with the neutral wire. Someone doing maintenance downstream could touch the bare ground and be electrocuted. Therefore, you should always use three-conductor cable—three insulated hot wires and a ground—for the dryer, such as 10-3 w/g NMB.

Another new code requirement is that the dryer must have a four-blade male plug and cord (also called a pigtail): two hot wires, one neutral and one ground. The female receptacle will have to match the pigtail (see the drawing on the facing page). The female receptacle can be surface-mounted or recessed. Surface mounts are more popular because they're easier to install. For the rough-in wiring, you must follow the guidelines listed in the NEC (electric clothes dryers are covered in sections 220.18 and 555.7e). It boils down to this: The design load must be at least 5,000 watts or the nameplate rating, whichever is higher. For a standard 240-volt electric dryer, that's 21 amps of current minimum, so the rough-in wiring must be 10 gauge, and the circuit will need a 30-amp breaker.

Heat Pumps

Heat pumps are normally wired by their installers, but sometimes they require the electrician to run the feed wires to the unit's location. If so, the installers will tell the electrician the size of wire they need for the system. There are normally two locations that need to be fed—one outside for the air handler, and one inside for the main unit. The wire I use most often for the outside air handler is 10-2 NMB w/g. The size of SE cable for the inside unit will depend mostly on the size of the electric backup heating. Disconnects are required at both the outside and inside unit if one is not already installed in the unit.

GARAGE-DOOR OPENER

All a garage-door opener needs is a standard receptacle. Although all other outlets in a garage must be GFCI protected, the outlet for the garage-door opener does not because it's normally located overhead, next to the opener itself. However, it is a good policy to use a GFCI here. People have been known to use the outlet for general-purpose use, even if they have to get on a ladder to reach it. In addition, a GFCI might help keep lightning off the opener's circuit and protect its electronics.

The opener must be installed exactly behind the end of the door as the door is opened and is normally mounted parallel to the floor. That means you can't install the rough-in wiring for the opener until you know the height of the garage door. For example, my garage has a ceiling height of 8 ft. and a 7-ft. garage door, so I mounted the opener 10 ft. back from the closed position or the door—and I power it from a GFCI receptacle mounted next to the opener.

SUBMERSIBLE PUMP

Submersible pumps come in many different horsepower ratings and in two different wiring configurations. The wiring to the well must be UL-approved for direct burial—it cannot be the standard yellow submersible-pump wire or even the twisted wire so commonly used within the well itself. The wiring within the well should be listed as pump wire. The gauge will depend on the horsepower of the pump and the distance to the well, as well as the depth of the well. A standard ½-hp pump, with its well 100 ft. from the house and a well depth of 140 ft., can get by with 12-gauge wire. A wire chart in the pump's instruction manual will list the gauges required.

It does not matter whether you want a two-wire pump with the starting circuits in the motor or a three-wire pump with the starting circuits in a pump control box. I don't have problems with either one. However, it is code to ground the pump and the well casing, if metal. For a two-wire pump, the branch-circuit wiring terminates at the pump's cutoff switch mounted immediately above the pressure switch. Standard two-conductor cable (with ground) can be used to connect the cutoff switch to the pressure switch. The wiring on the other side of the pressure switch will go straight to the well where the splice is made to the pump wire.

Wiring a Submersible Pump

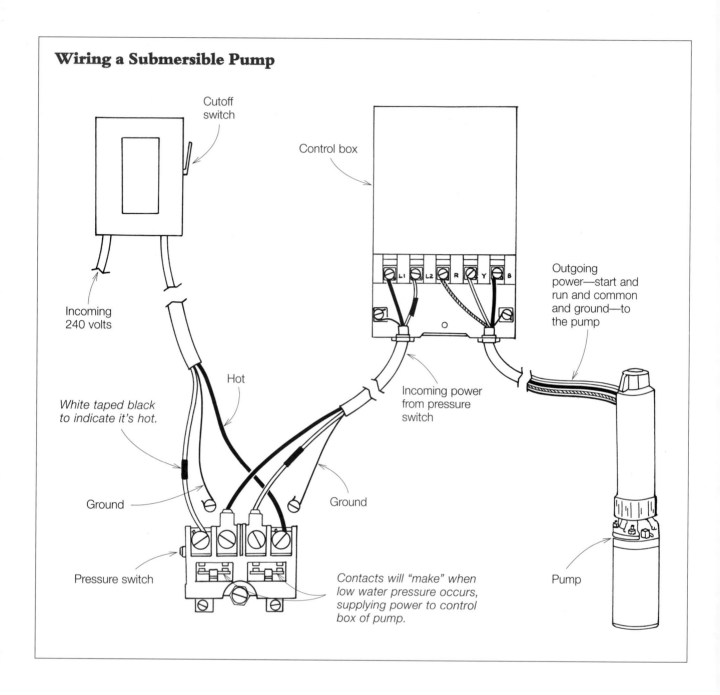

Cutoff switch

Control box

Incoming 240 volts

Hot

White taped black to indicate it's hot.

Ground

Ground

Pressure switch

Contacts will "make" when low water pressure occurs, supplying power to control box of pump.

Incoming power from pressure switch

Outgoing power—start and run and common and ground—to the pump

Pump

Wiring a three-wire (plus ground) submersible pump can get complicated (see the drawing at left). A cutoff switch is required for maintenance. Standard NMB cable is routed through the house, from the cutoff to the pressure switch. Each 240-volt leg connects to the two outer screws of the pressure switch. The two inner screws connect to the load—the pump control box—which is normally mounted within a few feet

of the pressure switch. Here the voltage is switched over to a start circuitry and a run circuitry to power the pump. The grounding wire must be kept all through the system—even down to the pump itself.

Do not match the breaker size to the wire run to the submersible pump. Many times a large-gauge wire is used to lower the resistance in the circuit on long runs to the well. For instance, for a ⅓-hp pump (a 4-amp load), do not install a 30-amp breaker just because you ran 10-gauge wire—use a 15-amp breaker. The smaller the breaker, the faster it will kick in case something happens.

WHIRLPOOL TUB AND SPA

Such items need to have their motors powered through GFCI-protected receptacles. Standard whirlpool tubs come with motors that are connected with a cord and plug. Bring the cable, normally NMB 12-2 w/g for whirlpools and much larger cables for spas, to the motor and then install a GFCI circuit breaker at the main panel. The receptacle and the motor itself are maintainable items and will need to remain accessible. Remember this because it may change the way the tub is to be oriented in the room. Small whirlpools will come with a built-in skirt and access panel. Larger units will have to have a skirt added on site and the access panel built in. Spas need GFCI protection as well. Sometimes the protection is included with the spa itself, and other times it is the responsibility of the installer. If it is the latter, you may have a problem. If the required GFCI breaker is of high amperage, the breaker may be hard to find and very expensive.

Always install a dedicated circuit for whirlpools and spas. If either unit is hard-wired, and a disconnect is not on the unit, a disconnect will be required. Also, for a spa, install an emergency cutoff switch and label it conspicuously. Note: An outside spa is a swimming pool, and rules involving pools, especially lighting, will apply (see NEC section 680).

INDEX

C

Cabinets, light fixtures for, 187
Cable fill:
  defined, 135
  and single-gang boxes, 168
  violation of, 135, 166, 169, 184
Cables:
  armor-clad, 12
  bundling of, 98
  channeling for, 102, 103
  defined, 7
  deciphering codes on, 24-26
  derating of, 98
  for electric dryers, 215
  fishing of, 105
  flexible, armored, 12
  identifying, 158
  metal-clad, 12
  nonmetallic, 13
  overheating of, 27
  protecting, 97
  running, 23, 27
  service entrance, 14
  sizing, 92
  stackers for, 101
  for submersible pumps, 215
  types of, 12-14
  and wire, 7
Channels, for cable, 102, 103
Circuit analyzers, use of, 44
Circuit breakers:
  double-pole,
    quad, 121, 122, 123
    ratings of, 121
    sizes of, 121-122
  installation of, 118
  limitations of, 122-123
  lockouts for, 201
  purpose of, 112, 117-118
  ratings of, 118-119
  single-pole,
    dual, 119-121
    full-size, 119-121
    half-size, 119-121
    ratings of, 119
    sizes of, 119

for submersible pumps, 216
  tabs for, 67
  two-part system in, 118
  wiring of, 118
  See also Main breaker
Circuit, defined, 5
Clamps, for ground wire, 81-82
Conductors:
  best, 9
  equipment-grounding, 74, 76
  grounded, 14
  grounding, 14
  grounding-electrode. *See* Ground
    wire
  noncurrent-carrying, 74
  See also Cable. Grounding. Wire
Conduit:
  grounding of, 161
  pulling wires through, 99
  See also Mast
Conduit benders, use of, 42
Cooktops, drop-in, 205
Covers,
  for outdoor outlets, 145.
  See also Boxes. Receptacles
Crawlspaces, reference point in,
  creating, 106
Current:
  calculating, 6
  decreasing, 6
  defined, 5
  increasing, 6
  measuring, 6
  and resistance, 6
  stray,
    protecting against, 76, 77
    sources of, 76
  types of, 5, 6
  See also Alternating Current. Direct
    Current
Current flow:
  defined, 5
  discussed, 14
Cutoff panels:
  defined, 48, 60
  use of, 48
  when required, 60
  See also Subpanels. Main panels
Cutters:
  diagonal, 30
  side, 30

D

Derating, defined, 70
Dimmers:
  bulbs for, 180
  and buzzing, 180
  components of, 177, 178
  controls for, 177
  for fluorescent fixtures, 178
  heat sinks for, 180
  and noise, 179
  overloading of, 180
  problems with, 179-180
  types of, 177
  use of, 177-178
  wiring of, 178, 180
Dining rooms:
  receptacle layout in, 93
  switch layout in, 93
Direct current:
  defined, 4
  generating, 5, 15-16
Dishwashers:
  voltage for, 201
  wiring of, 201-202, 203
Drill bits:
  auger,
    diameters of, 36
    problems with, 35
    use of, 35, 36
  extension, 41
  hole saws,
    types of, 36-37
    use of, 36-37
  spade,
    problems with, 35
    use of, 35
  stepped,
    design for, 41
    use of, 41
  twist drills,
    types of, 37
    use of, 37
  wood borers, self-feed, 36
Drills:
  cordless,
    advantages of, 39
    chucks for, 39
    extension bits for, 41
    use of, 39

pistol-grip, 34
right-angle, 34
for roughing-in wiring, 33
use of, 33-34
See also Drill bits
Dryers, electric:
cables for, 215
circuit breakers for, 122
plugs for, 215
receptacles for, 214-215
wiring of, 213-215

E

Electrical tape:
for insulating terminals, 139
splicing with, 32
types of, 32
uses of, 32
Electricity:
elements of, 5-7
flow of, 4, 5
See also Current. Ohm's law.
Resistance. Voltage
Exothermic weld, for ground wire,
82-83
Extension cords:
buying guidelines for, 33
gauges of, 33
GFCI-protected, 33
types of, 33

F

Fans, ceiling:
boxes for, 190
buying guidelines for, 187-189
clearances for, 187
controls for, 181
cooling with, 188, 189-190
custom, 189
and humming, 188
installing, 189-190
lights with, 191
noise with, 181
reversible, 190-191
sizing, 188

supporting, 189
warming with, 188, 189-190
wiring of, 191
Fans, exhaust, for stoves, 206-207
Fish tapes:
characteristics of, 42
homemade, 42
for pulling wire, 42
store-bought, 42
use of, 42
Fish wire, use of, 105
Fuse boxes:
recognizing, 117
replacing, 117
Fuses:
cartridge,
ferrule-type, 113
knife-blade, 113
materials in, 113
problems with, 113, 114
testing, 114
development of, 113
limitations of, 122-123
plug,
adapters for, 116
for appliances, 115-116
limitations of, 114
materials in, 114
modifying, 117
problems with, 116
sizes of, 114, 116
testing, 114, 116-117
time-delay, 114-115
point-of-use, 116
purpose of, 112

G

Garage-door openers:
locations for, 215
receptacles for, 215
Garages:
receptacle layout in, 94
switch layout in, 94
Garbage disposals:
switch for, 203
wiring of, 202, 203
Gauge, *See* Wire

GFCI, *See* Ground-fault circuit
interrupters
Ground:
resistance of, 79
testing for, 162
Ground fault:
defined, 74, 112
direct, 74, 76
on grounded appliance, 75, 76
indirect, 76
occurrence of, 74-76
preventing, 74, 76
on ungrounded appliance, 75, 76
Ground rods:
driving, 80
installing, 79-80
locating, 79-81
material for, 79
number of required, 79-80
purpose of, 79
size of, 79
Ground wire:
attaching to ground rod, 83
clamps for, 81-82
defined, 80
exothermic weld for, 82-83
gauge of, 80-81
purpose of, 80
running, 81
sizing, 81
splicing, 81
Ground-fault circuit interrupters:
in basements, 94
in bathrooms, 93, 129
circuit-breaker,
discussed, 126-127
ratings of, 127
wire for, 132
wiring of, 132
defined, 124
exceptions, 130
for extension cords, 33
fooling of, 126
in garages, 94, 129
and grounding, 130
grounding of, 131
and hot tubs, 130
in kitchens, 92, 129
outdoors, 129-130

W

PUBLISHER: **Jon Miller**

ACQUISITIONS EDITOR: **Julie Trelstad**

EDITORIAL ASSISTANT: **Karen Liljedahl**

EDITOR: **Thomas C. McKenna**

DESIGNER/LAYOUT ARTIST: **Henry Roth**

PHOTOGRAPHER, EXCEPT WHERE NOTED: **Rex Cauldwell**

ILLUSTRATOR: **Ruth H. Steinberger**

TYPEFACE: **Stone Serif**